GREEN $ENSE
FOR THE HOME

Rating the **Real Payoff** from

50 Green
Home Projects

Eric Corey Freed
and Kevin Daum

The Taunton Press

The Taunton Press
Inspiration for hands-on living®

The Taunton Press, Inc., 63 South Main Street, PO Box 5506, Newtown, CT 06470-5506
e-mail: tp@taunton.com

Editor: Courtney Jordan
Copy editor: Diane Sinitsky
Indexer: Jay Kreider
Cover design: Jean-Marc Troadec
Interior design: Alison Wilkes and Susan Fazekas
Layout: Susan Fazekas
Illustrator: Mario Ferro
Cover/Graphic Icon illustrator: Lucy Vigrass

Library of Congress Cataloging-in-Publication Data

Freed, Eric Corey.
  Green sense for the home : rating the real payoff from 50 green home projects / Eric Corey Freed and Kevin Daum.
     p. cm.
  The letter S in the title is represented by a dollar sign.
  Includes index.
  ISBN 978-1-60085-155-1
  1. Dwellings--Maintenance and repair--Amateurs' manuals. 2. Dwellings--Energy conservation--Amateurs' manuals. 3. Energy consumption--Cost control--Amateurs' manuals. I. Daum, Kevin. II. Title. III. Title: Green $ense for the home.
  TH4817.3.F74 2010
  643'.7--dc22

                          2010000308

Printed in the United States of America
10 9 8 7 6 5 4 3 2 1

The following manufacturers/names appearing in *Green $ense for the Home* are trademarks: Aged Woods®, Agriboard®, Amana®; Amazon.com®, American Hydrotech, Inc.®,Antron®, AquaHelix™, AquaTimber™, AQUS®, Armstrong®, Avonite™, BioBased Insulation®, BioTimber™, Blomus®, BottleStone™, Breathe Easy®, Brita®; Brondell® PerfectFlush™, Bubble Wrap®, Built Green®, Cali Bamboo®, ChoiceDek®, Corian®, CorrectDeck®, Cradle to Cradle℠; CraftWood®, Danby®, Durisol®, Earth911®, EarthSource Forest Products®, Ecoboard®, EcoRock®, EcoSmart®, EcoTimber®, EcoTop™, Energy Star®, EnviroGLAS®, Envirosink®, Equator®, Eternal Hybrid™, EverGrain®, Evolve®, Expanko®, FireCrystals®, FischerSIPS®, Formica®, Fresh Aire Choice®, Gaiam®, The Garland Company, Inc. ®, Green Seal®, GreenFiber™, GreenGrid®, Greenline™, GreenTech®, Habitat for Humanity®, Heliodyne™, The Home Depot®, Humabuilt®, IceStone®, IKEA®, InterfaceFLOR™, Java-Log®, KitchenAid®, Jenn-Air®, Klean Kanteen™, Lego®, Lowe's®, Lutron®, Marmoleum®, Maytag®, MicroFridge®, Nalgene®, NEFF Kitchens®, Nova Distinctive Floors™, NUDURA®, Nu-Wool®, Olympic®, Oxygenics®, Pallas®, PaperStone™, Perma-Deck®, PlasTEAK®, Pur®, PureBond®, ReWater® Systems, RhinoDeck®, Ritz-Carlton®, Richlite®, Rumber®, Safecoat®, Safretti®, Shaw Contract Group®, Sheetrock®, Sigg™, Silestone®, Solatube® Solar Star®, Spider®, Styrofoam™, SunPower®, SunTouch®, Surface iQ™, Teflon®, Teragren®, TimberTech®, TiVo™, Trestlewood®, Trex®, UltraGlas®, Uni-Solar®, USFloors®, Vetrazzo®, Warmboard®, WarmWire™, WaterSaver Technologies™, WaterSense®, WeatherTRAK®, Whirlpool®, Wilsonart®, YOLO Colorhouse®, Zodiaq®

About Your Safety: Homebuilding is inherently dangerous. From accidents with power tools to falls from ladders, scaffolds, and roofs, builders risk serious injury and even death. We try to promote safe work habits throughout this book. But what is safe for one person under certain circumstances may not be safe for you under different circumstances. So don't try anything you learn about here (or elsewhere) unless you're certain that it is safe for you. Please be careful.

© **Mixed Sources**

Product group from well-managed forests, controlled sources and recycled wood or fiber
www.fsc.org Cert no. **SW-COC-002550**
© 1996 Forest Stewardship Council

FSC

*To my beautiful daughter, Grayson, whose birth coincided with
the birth of the idea for this book.*
—E.C.F.

*To every consumer who wants to be green without suffering
the financial losses of ignorance or confusion.*
—K.D.

Acknowledgments

Green$ense for the Home was written to provide you with the research
and arguments you need to make an informed decision on greening
your home. We do not rate or make recommendations regarding specific manufacturers or trade names. Any manufacturers mentioned are
for your reference only.

Eric thanks the dedicated interns and researchers who contributed
their time and talents—Geoffrey Allen, Megan Bierwirth, Katie Eberle,
Cristina Foung, Graham Grilli, Meredith Hart, Soojin Hur, Elisa Kim,
Erika Ohlsson, Consuelo Pierrepont, Owen Schoppe, Lynette Scott, and
Shayan Shagari—and Matt Golden of Sustainable Spaces for his clear
and practical insight.

Kevin thanks Robin Katz for her technical/financial research and
support; Carolyn Roark and Deanna Daum for their efforts on photo
collection; and Bradford Rand of Go Green Expo for inspiring a make-
sense green approach.

Together, the authors thank Peter Economy for his coaching, editing,
and guidance; The Taunton Press team for their belief and support; and
Joy Tutela of The David Black Agency for her persistence and support.

contents

· ·

13 green
home projects
you can do
when building new

introduction

If we all believe saving the planet is critical, then rating green home projects from a practical point of view might seem counterproductive. In a perfect world, we would do everything possible to save the planet and not have to pick and choose among worthwhile projects. Unfortunately, greening a home gets expensive, and good intentions can go by the wayside when the time comes to pull out your wallet. And while some projects may be worth making financial sacrifice, most of us haven't had access to the information necessary to assess the true affordability of green improvements until now.

Green$ense for the Home provides in-depth analysis of 50 green home projects from two experts with quite different perspectives. Eric represents the Greenies of the world. As the nation's leading green architect, he is in the pantheon of environmental do-gooders for his "always do the green thing" approach to home design. Eric shows why a project is good for a home and the environment. He explains how to accomplish the project and gives you valuable expert advice and resources for getting it done right.

On the other side, Kevin represents the hard reality of practicality and affordability. He examines how much a project will cost or save in plain dollars and cents, and also assesses practical pitfalls for each project. After financing more than 1,000 custom homes, Kevin has seen mistakes galore where green aims went horribly wrong. Kevin is a strong green supporter, but he provides perspective about what is affordable and desirable in the housing market.

The good news is that there are many green home projects that can be accomplished with minimal cash investment. And many will actually pay you back—and keep putting money in your piggy bank each and every day. Together, we give you a realistic, balanced view of the pros and cons of each project so you can make green choices that best serve your planet, your wallet, and your lifestyle. For example, if Eric were to tell you about all the green reasons why you should use recycled toilet paper to save old-growth forests, Kevin would let you know that even though you could save $0.50 per 100 sq. ft. by buying earth-friendly TP, he and many others are not going there until it is made squeezably soft!

ERIC SAYS
Eric represents the Greenies of the world.

KEVIN SAYS
Kevin represents the hard reality of practicality and affordability.

All of the projects in this book are here because they will improve your home and help our environment. We identify the global benefits for each project, but our main goal is to provide you with a way of figuring out the projects that will work for you financially, while providing practical advice on how to best implement them. Whether you are building a custom home from scratch, remodeling an existing home, or adding a small green improvement, analyzing the resultant cost, savings, and value added is of paramount importance when making the decision whether to take on a green project, and we put all that key information in your hands.

How to Use This Book

This book is built in three sections based on the amount of time each project will likely take to accomplish:

- **16 Green Home Projects You Can Do Today** covers green projects that can be done immediately.
- **21 Green Home Projects You Can Do Tomorrow** discusses projects to do when undertaking a remodel or when you've got extra time or money to invest in the home.
- **13 Green Home Projects You Can Do When Building New** includes projects for new, custom homes that involve home design and extensive construction decisions.

At the start of each project, you'll find **Green $pecs,** a ratings sidebar that uses handy reference icons to clue you in to crucial information about what you can expect from each project:

- **Overall Rating** The overall rating balances the green and financial aspects of each project to give you a sense of how worthwhile it is. The more icons, the better the project's value.
- **Difficulty** Eric's assessment of how difficult each project is. The more hammers you see, the more involved the project is.
- **Green Benefits** This is a list of the environmental gains that can be achieved from each particular project. The categories include Energy Saving, Global Warming Reduction, Resource Conservation, Water Saving, and Health Improving.

Green $pecs

Overall Rating

Difficulty

Green Benefits

Overall Rating We've weighed the green and financial aspects of every project to come up with an overall rating that indicates how sensible and useful each one is for you. The icon shows the two sides of our debate in the form of a section of the globe topped with a dollar sign. The rating is 1 to 5, with the higher number of icons meaning the project has a higher value.

Difficulty Home projects come in many different levels of expertise. For some, you'll need a few minutes and your own two hands. For others, you'll need to call in reinforcements. Many projects fall somewhere in between. Look to this icon to see the level of involvement and the skill (on a scale of 1 to 5) you'll need to complete the job.

Energy Saving This icon indicates that a project has a positive impact on the amount of energy used in a household. The project helps with energy efficiency and may result in utility bills—from electricity to gas and heating—going down.

Global Warming Reducing Many of our projects have an impact on one of the biggest environmental threats of our time—global warming. Projects with this icon reduce that threat by decreasing the main causes of global warming, including deforestation, carbon dioxide emissions, and the burning of fossil fuels.

Resource Conservation Reducing household waste, reusing and recycling products and materials in the home, and buying items that contain recycled content are all ways to conserve resources. Projects with this icon play a part in getting this done.

Water Saving Water is one of our most precious resources but becomes more and more endangered every day. This icon means a project helps reduce water consumption and, in some cases, helps recycle water for reuse in the home.

Health Improving It's nerve-wracking to think that materials and items in our homes could be doing us harm, but sometimes that is the case. If a project has this icon, it means that air quality within a home is improved by reducing the amount of noxious fumes and solvents in the house, or preventing them from coming in altogether.

Resources for materials, contractors, and financial information are provided at the end of each project.

ERIC SAYS Eric starts off each chapter by introducing the project and its green potential. He assesses the project in terms of the following questions:

- What will this project do for your home?
- What will this project do for the Earth?
- Will you need a contractor?
- What are the best sources for materials?
- How much maintenance will be required after installation?
- How long will the project take to accomplish?

Along the way, Eric provides helpful illustrations, step-by-step photographs, and resources, and makes sure you have everything you need to successfully take this green home project from start to finish.

KEVIN SAYS Kevin addresses the practical pros and cons of the project. Eric may roll his eyes at Kevin's cold fiscal concerns, but Kevin investigates the validity of the green claims made in each project. Just remember, however, as Eric points out, that even if there isn't any recorded data it doesn't mean there's no green value. Kevin also works through the financial issues, so you can make an informed decision based on these considerations:

- **What is the capital cost?**
  Here you'll find information on the upfront cost of the project. The more red dollar signs (on a scale of 1 to 5) there are, the higher the cost. Depending on the nature of the project, it may be based upon the entire cost or it may be calculated on a per-sq.-ft. basis.
- **What financial resources are available?**
  This is where you'll find information on ways to pay for each project. Some projects may be eligible for rebates or tax credits. Some may be easily financed. The more benefits available, the more green dollar signs (on a scale of 1 to 5).
- **What is the monthly cost or savings?**
  This section addresses how a project impacts a monthly housing budget. Not every green project is a money-saving proposition. If the technology hasn't advanced far enough yet to have broad commercial appeal, you may pay more for the privilege of going green. Yes, it's hard to put a price on saving the planet, but since that's the crux of the book we feel obligated to share the facts. Depending on the project, this section could show net savings or costs. The more

savings possible, the more green dollar signs are used. The higher the cost, the more red dollar signs you will see. Both red and green dollar signs are on a scale of 1 to 5.

- **What is the long-term home value?**

  There are two major factors taken into consideration when figuring out the long-term impact on home value. First is marketability— will someone pay more for a house because of a particular green improvement? How will a bank appraiser factor in the improvement? The second factor is obsolescence. We are on the cutting edge of green technology and advances are made daily. As more money is pumped into research, improvements will be made and the project that requires thousands of dollars may be outdated in a few short months or years. In this section, we evaluate the technology and how easy it is to upgrade it combined with marketability. The more green dollar signs you see (on a scale of 1 to 5), the more benefits you could get. (For you renters, showing this info to your landlord might get him or her to cough up some dough for improvements. It never hurts to ask!)

THE BOTTOM LINE IS...

Together, we want to empower you to make green choices based on the affordability of the projects and the potential cost savings, with full knowledge of the ethical and financial benefits of each. Long nights of debate and discussion between the two of us (under compact fluorescent lights, of course) went into determining an overall assessment that is both legitimate and fair. Some factors will change over time, and we look forward to your feedback from your own experiences, which you can share freely at www.greensensebook.com, but we've done our best to give you the information you'll need to decide what green way is right for you. Okay—enough talk. Go get some green sense and start saving!

16 green home projects you can do today

Whether you rent or own your home, these simple projects will help to quickly reduce energy and water usage. All of the projects can be completed in just a day or two, and most are things you can do yourself without any experience.

Change your light bulbs

ERIC SAYS How many environmentalists does it take to change a light bulb? There are several answers to this joke (none of them that funny), but the correct answer is: "all of them." Everyone should swap burned-out incandescent bulbs with energy-efficient compact fluorescent (CFL) ones because upgrading light bulbs is one of the simplest but most effective things you can do to save energy in your home.

As a nation, we spend more than $37 billion annually just on lighting. In an average home, lighting accounts for 20% to 30% of all electricity use. (At work, that figure goes up to almost 40%.) A CFL uses only a quarter of the energy of an incandescent bulb of the same brightness and lasts much longer, meaning less waste and less changing out bulbs.

What will this project do for your home?

Traditional bulbs give off 90% heat and only 10% light. A CFL bulb is almost the opposite. An 18-watt CFL provides the same brightness as a 100-watt incandescent bulb, performing the same job and using a fraction of the energy. And the life of a CFL is up to 10,000 hours per bulb compared with less than 1,300 hours for a typical incandescent.

What will this project do for the Earth?

Traditional incandescent bulbs waste energy, give off a lot of heat, burn out quickly, and end up in the trash (more than 5.5 million bulbs are thrown away every day)—all of which adds up to more global warming, higher air-conditioning costs, and more waste in already crowded landfills.

Compact fluorescent (CFL) bulbs have a distinctive shape but screw into any standard bulb socket.

If every American home replaced just one light bulb with a CFL bulb, we would save enough energy to light more than 2.5 million homes for a year and prevent greenhouse gases equivalent to the emissions of nearly 1.2 million cars. If every American household took the additional step of replacing their most-often-used incandescent bulbs with CFLs, electricity use for lighting in the United States could be cut in half.

Will you need a contractor?

No. CFL bulbs fit into a standard light bulb socket—they work with your existing light fixtures and do not require any special tools or skills. Switching to CFLs is truly as simple as screwing in a light bulb.

What are the best sources for materials?

Compact fluorescents are now available in a variety of shapes and sizes everywhere light bulbs are sold, so you should have no problem finding bulbs to fit any fixture in your home. If you have a unique lamp or fixture, take the existing bulb along to the store to compare sizes. Dimmable CFLs are also available. They cost a little more but will work with existing dimmer switches, whereas a regular CFL will only light up if the dimmer is at the highest setting. CFLs can be used with timers, motion detectors, photocells, and occupancy sensors.

When CFL bulbs were introduced in the early 1990s, the quality of the light was a big issue—early bulbs gave off light that was cold, antiseptic, and hard to look at—despite the long-lasting energy efficiency of the bulbs. In the last two decades, CFL technology has improved greatly. New color temperatures and warmer bulbs have been introduced, and now the quality is the same or better than that of a traditional bulb.

All light bulbs have a "color temperature" indicating the warmth of the light. Warmer bulbs are yellowish in color, whereas cooler ones are bluer. Mixing and matching bulbs from different manufacturers can

Mercury Emissions of CFLs and Incandescent Bulbs

Each CFL bulb contains 4 milligrams of mercury and emits 2.4 mg during the manufacturing process. But even combined, those two amounts do not equal the 10 mg of mercury needed to produce an incandescent bulb.

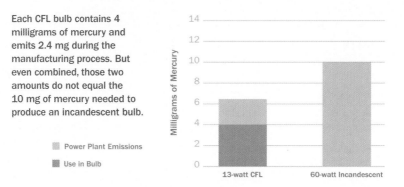

Power Plant Emissions

Use in Bulb

(Adapted from Energy Star, Department of Energy)

create problems because they may be slightly different colors and not match precisely. To avoid this, buy several cartons of bulbs at a time rather than just a few as you need them.

If you're worried about the buzzing and flickering problems commonly associated with fluorescent tube lights, don't be. Modern compact fluorescent bulbs use special electronic ballasts to eliminate those issues. Most CFLs are rated for outdoor use, but read the package carefully to make sure if you are installing one in an exterior light.

How much maintenance will be required after installation?

There is concern about the mercury trapped inside CFLs, but the small amount of mercury in a CFL—4 milligrams (mg)—is much less than the 10 mg of mercury released into the air by using an incandescent bulb, which requires more coal to be burned (and even more mercury released into the air) to generate the additional electricity required over its lifetime. The average amount of mercury in CFLs has dropped 20% in the past year, and manufacturers continue to reduce the mercury content in the bulbs they produce.

But CFLs must be recycled, not thrown in the trash. In fact, many states prohibit disposing of any type of fluorescent lamp in the solid-waste stream. Recycling the lamps is simple: Call 800-CLEAN-UP (800-253-2687) or visit the Earth 911 website (www.earth911.com) and enter your zip code to find the nearest recycling center.

How long will the project take to accomplish?

Swap out every light bulb in your home to capture the maximum energy savings or do it gradually as incandescent bulbs burn out. Either way, it shouldn't take you more than a minute or so per bulb.

KEVIN SAYS Lighting is a large part of home energy use, and Greenies are going after it in a big way. Laws are being suggested or are already on the books in several states to stop use of Thomas Edison's incandescent light bulbs entirely. And technologies such as light-emitting diodes (LED) and Nikola Tesla's inductive (magnetic) lighting are gaining popularity on the CFL-type bulb. They have a ways to go before the cost makes them attractive for everyday use, but expect to see LEDs on a large scale within five to seven years, possibly replacing CFLs completely.

The biggest complaint about CFLs is the cold, harsh feel of the light, which is slowly improving. *Popular Mechanics* did a study on lighting that showed people preferred quality CFL lighting to incandescent bulbs, but many still use incandescent bulbs for their warm tone.

Then there is the mercury concern. If you break a CFL, you release a small amount of harmful gas into the atmosphere, and when the bulbs burn out, you can't just toss them in the trash.

What is the capital cost? $

Obviously, most light bulbs aren't very expensive. Even though a CFL bulb might be three to four times the cost of a normal incandescent bulb, it is still affordable and prices are coming down. An eight-pack of 15-watt CFL bulbs can be found online for $10 to $12 compared with $5 for 60-watt incandescent bulbs. You could incur additional cost if the aesthetics of built-in fixtures require specialized bulbs.

What financial resources are available? $$$$

As with most early adopted Green technologies, light bulbs have government support across the country. Energy Star® offers rebates for as much as $2 through local government agencies and local stores. A quick online search for "CFL rebates" will yield many coupons and offerings.

Converting to CFLs

When buying CFL bulbs, select wattage of only about one-fourth of what you usually buy.

25-watt incandescent = 7-watt compact fluorescent

40-watt incandescent = 11-watt compact fluorescent

60-watt incandescent = 15-watt compact fluorescent

75+-watt incandescent = 18-watt compact fluorescent

RESOURCES

Earth911
www.earth911.com
1-800-CLEAN-UP

Energy Star
www.energystar.gov

Flex Your Power
www.fypower.org

My EnergyStar
www.myenergystar.com

Also check with your local power company or go to www.energystar.gov to find bulk suppliers. Local university or conservation groups often have free light bulb swaps. Even retailers like The Home Depot® and IKEA® periodically swap bulbs.

What is the monthly cost or savings? $$$$$

Monthly savings for energy-efficient lighting are all over the map, depending on the wattage you buy, how the light is used, and the cost of energy in your area. The accepted rule of thumb is roughly a 75% reduction in energy usage for comparable lighting.

How long it takes to recoup costs varies widely based upon the quality of lights, the price paid, and how often you use them. Assuming conservatively that a CFL bulb lasts eight times longer than an incandescent bulb and costs three times more, the cost of the bulb is quickly recouped and you pick up the energy savings as a bonus. Getting rebates will put you way ahead of the game.

What is the long-term home value? $

The impact on home value is nearly nonexistent when it comes to changing light bulbs. The minimal impact on value stems from built-in fixtures. While manufacturers have successfully adapted CFLs to work with dimmers and fit into sockets, many fixtures are designed around the shape and aesthetic of incandescent bulbs. If you invest in new fixtures with CFL technology, you may find it eventually replaced by advances in LED or induction lighting.

When selling your home, you may need to modernize by changing or upgrading fixtures. At the very least, you will need to invest in bulbs of the right tone and color to help make the house look warm and inviting.

THE BOTTOM LINE is...

It's true the light of CFLs can take some getting used to, but the quality is steadily improving. Changing light bulbs equals free money in your pocket and definitely helps the environment. Don't let the CFL mercury complaints deter you either, since the levels are far lower than those found in sushi. Just dispose of CFLs properly when they burn out. That will keep the mercury out of the dumps and away from the water in which your soon-to-be spicy tuna roll may be swimming. ∎

Help your toilet use less water

ERIC SAYS In the United States, more water is consumed per person than in any other country. One reason is that our homes consume an incredible amount of fresh drinking water, and most of it is used in places where it does not need to be so, well, drinkable. More than a quarter of all of the water used inside the home is flushed down the toilet, which is literally a waste. The toilet is the single largest user of clean drinking water in the home, and it is also the easiest place to conserve water.

Toilets manufactured prior to 1990 use an average of 3.5 gal. (some up to 7 gal.) of water for every single flush. Newer toilets introduced in the 1990s, referred to as "low-flow" toilets, use only 1.6 gal. per flush and cost the same to buy and install as a water-wasting model. Low-flow toilets are required in California, but most states do not yet have these regulations. Before you run out and replace your existing toilets, there are a number of simple and effective things you can do to force

Green $pecs

Overall Rating

Difficulty

Green Benefits

Water Use in America and the Rest of the World

American water use is way above global averages. If you add up the water use of an average person in Brazil, Germany, the United Kingdom, China, and Honduras, it is still less than the amount used by the average American.

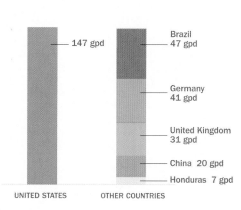

147 gpd — UNITED STATES

OTHER COUNTRIES:
- Brazil 47 gpd
- Germany 41 gpd
- United Kingdom 31 gpd
- China 20 gpd
- Honduras 7 gpd

(Adapted from data courtesy of the Environmental Protection Agency)

your old toilet to use less water, from installing a flush adapter to making flusher adjustments and installing an Aqua® system, which reuses soapy water to flush the toilet. And when the time comes to replace your working toilets, make sure you buy a low-flow model.

What will this project do for your home?

Nearly 55 billion gal. of clean, drinkable water is flushed down toilets across the United States each day. Up to half of that amount of water can be saved by using simple water conservation measures, and a one-time retrofit of your toilet will save thousands of gallons of water in your home every year.

What will this project do for the Earth?

A worldwide water crisis is looming. In the future, water that was once inexpensive will become unaffordable to many. Making the right choices now will help protect everyone in the near future.

Will you need a contractor?

There are a few water-saving options for existing toilets that you can do yourself as well as an inexpensive weekend project that may require the help of a plumber.

Lowering the toilet tank float manually When you flush the toilet, the water in the tank rushes into the bowl and washes away waste. A float in the tank drops as the water in the tank drops. This

Household Indoor Water Use Per Year

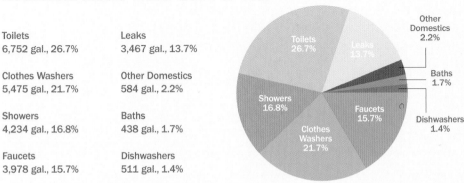

Toilets
6,752 gal., 26.7%

Leaks
3,467 gal., 13.7%

Clothes Washers
5,475 gal., 21.7%

Other Domestics
584 gal., 2.2%

Showers
4,234 gal., 16.8%

Baths
438 gal., 1.7%

Faucets
3,978 gal., 15.7%

Dishwashers
511 gal., 1.4%

(Adapted from data courtesy of the Environmental Protection Agency)

Flush with Less Water

In the tank of your toilet, there is a float. After flushing, water fills the tank and the float rises until it reaches the set level. By lowering this float, you are tricking the toilet into using less water to flush.

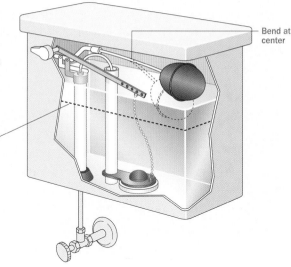

Bend at center

New water level

activates a valve that refills the tank with fresh water. As the water rises in the tank, the float rises with it. Once the water level reaches the set height, the valve shuts off.

There are two types of floats. Older floats are held in place by a metal arm attached with an adjustment screw. Newer floats move up and down along a shaft. By lowering the height of the float, you reduce the amount of water that fills the tank. You'll need a screwdriver (or strong hands), the nerve to stick your hands into the back of the toilet, and just a few minutes.

With a screwdriver, slowly turn the adjustment screw to lower either type of float. If you don't have a screwdriver handy, you can gently bend the float arm in the center to lower the float. This is a less precise, but handy, alternative.

That's it. By lowering the float level, you have tricked the toilet into filling with less water. You can save up to 2 gal. of water per flush with this method. Try different heights to find the amount of water that works for you.

A word of caution: Bending the float arm could hinder the normal operation of the toilet. There is a slight risk you could break the toilet and need rescuing by a real plumber. Only proceed if you're comfortable doing this work.

Placing a beverage bottle in the back of the tank You'll need an empty 2-liter soda bottle; sand, marbles, or small pebbles; and about 5 minutes.

A soda bottle filled with stones (right) and inserted into the toilet tank (left) can cut the amount of water your toilet uses by a half gallon per flush.

Wash out the bottle and fill it about half to three-quarters of the way with sand, marbles, or small pebbles. Fill it the rest of the way with water and tightly seal the cap.

Slip the bottle into the toilet tank as far from the valve in the bottom as possible. Make sure it can stand up by itself. Replace the tank lid. The bottle takes up some of the volume of the tank and will save half a gallon on every flush and thousands of gallons of drinking water a year.

Installing low- and dual-flush adapters Toilets flush two types of waste: liquid waste and solid waste. Yet regardless of the type of waste, the same amount of water is used to flush the toilet and, 90% of the time, water is wasted.

While the average toilet uses 3.5 gal. for every flush, a "low-flow" toilet uses 1.6 gal. per flush, and an "ultra-low-flow" requires only 1.1 gal. But a clever alternative is a "dual-flush" toilet. Two flush buttons are located on the toilet. Push one button and the dual-flush uses only 0.8 gal. per flush for liquid waste. For solid waste, push the second button and the full 1.6 gal. is used.

Rather than replace your entire toilet, a conversion kit can transform your toilet into a dual-flush model. Products like a dual-flush adapter install simply by unscrewing a toilet's existing flush handle and inserting the new handle.

What are the best sources for materials?

All of the products mentioned here can be ordered online direct from the manufacturer. You may also find them in your local hardware store, but it may help to call first. You can find a list of retailers at the manufacturers' respective websites.

You may be concerned that there is a health risk in using these measures in your toilet. Not to worry—you're working on the clean water side of things. None of these installations pose any sort of health issue.

How much maintenance will be required after installation?

All of these toilet conversions are virtually maintenance free. You'll see an immediate and noticeable drop in your water bill. Check your usage before and after the installations to see the full impact.

KEVIN SAYS Growing up in Los Angeles in the '70s, I remember putting bricks in toilets during droughts. Now with climate change threatening to bring drought conditions to everyone, I can certainly see the value of water conservation across the board.

The biggest issue with toilets is flushing power. Luckily, my new hero, engineer Bill Gauley of Veritec Consulting, created what is now the standard flush test, which shows that water volume is only one factor in flushability. Apparently, simply reducing water in some toilets doesn't work effectively due to bowl shape.

Pressurized retrofit systems now widely used in hotels seem to solve the bowl clearance issues. But the best conservation solution takes advantage of the roughly 5 to 1 ratio of going number 1 to going number 2. The dual toilet allows for lower water usage for most visits and gives extra flow when needed. And dual-flush retrofits mean you don't have to replace the whole toilet.

What is the capital cost? $$

Prices for toilet conversions vary greatly. Eric's soda bottle suggestion will cost you $1.25 and, of course, you get to drink two liters of your favorite soda. A dual-flush retrofit will cost roughly $50 per toilet, which looks to be the best deal all around. If you want to go the permanent route, new low-flow or pressurized systems range from $350 to $750 plus installation. We'll leave the luxury toilets off the rating since at $5,000+ they are strictly for the Trump crowd.

A dual-flush adapter converts any toilet into a water-efficient model by using a special flush valve to control the amount of water used. The control pad has two buttons to choose between for a full or half flush. Installation is straightforward but might require the help of a plumber if you're not that handy.

Recycle Your Water

After water is used in a sink, shower, or washing machine, it is called "gray water." The term refers to the grayish cast of the used, soapy water. Most of the time, gray water simply goes down the drain. But you can recycle gray water and use it to flush your toilet. After all, the water in the toilet does not need to be drinkable. By flushing a toilet with gray water, you can reduce your water bills by more than 25%.

The AQUS System shown below collects soapy water from the sink and pumps it to the adjacent toilet. Installation is fairly complicated and only for those comfortable with plumbing work. A plumber should be able to install the unit in less than an hour.

Once installed, systems like the AQUS conserve virtually all of the water flushed down the toilet. Although the AQUS filters and cleans the water, you'll need to clean your toilet bowl regularly or soap scum will build up.

The AQUS requires simple annual maintenance of cleaning the filter and replacing the disinfecting tablets, which control bacteria and other contaminants. This takes less than 15 minutes. Otherwise, the system is maintenance free. If you don't use the sink, the toilet will simply use fresh water to fill the tank.

A system like this is best installed on only your most-used toilets. It isn't as efficient for a guest bath or a toilet that is rarely used.

How the AQUS System Works

The AQUS System stores soapy water from the sink and runs it to the toilet, flushing it with recycled water.

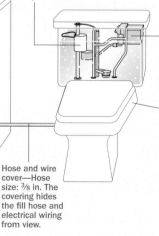

Tablet dispenser—Reused water is rinsed over disinfecting tablets. Makes reused water safe for people and pets. But the water is still nonpotable water—do not drink.

Fill valve—AQUS does not cross connect to fresh water or interfere with backflow prevention.

Water-control unit (WCU)—Patented water-control unit delivers treated, reused water as the primary source for flushing toilet.

Lavatory tubing—System connects via standard PVC tubing. Parts included.

Toilet bolts—Patented bolts allow reused water to enter from the bottom of the toilet tank.

Pump—12 Volt DC pump. Average flow rate: 1.6 gal./72 secs.

Screen filter—Filters large particles such as hair, toothpaste, etc.

Water reservoir—Size: 13.5 in. × 15.5 in. × 8 in. Volume: 5.5 gal.

Hose and wire cover—Hose size: $3/8$ in. The covering hides the fill hose and electrical wiring from view.

(Image adapted from WaterSaver Technologies, www.watersavertech.com)

What financial resources are available? $$$

Happily, there is a wide array of toilet rebates available that will cover about half your cost. They vary based on a home's local water utility and the product purchased, ranging from $30 rebates for dual retrofits to $300 for full toilet installs. Check out http://epa.gov/watersense/pp/find_rebate.htm or www.toiletrebate.com to find programs.

What is the monthly cost or savings? $$

Unless you are on a well system, toilet usage comprises roughly 25% of a household's water bill. If the typical ratio of going number 1 to number 2 is 5:1, then it stands to reason that installing a dual-flush adapter will reduce your toilet water usage by 40%. That translates into approximately 8% monthly savings, or $4 on a $50 monthly bill. That's not enough to cover a new toilet in less than five years, but it will go a long way toward the retrofit and some extra toilet paper every month.

What is the long-term home value? $$$$$

Research shows that the two best places to invest your home-improvement dollars are kitchens and bathrooms. People will pay more for bathrooms that are state of the art and will pay less when they think they will have to retrofit or remodel. Since many states require low-flow retrofitting at sale, you'll end up coughing up the money at closing if you don't make the toilet fix in advance, so it's in your interest to make the change and enjoy the conservation benefits.

THE BOTTOM LINE IS...

Although toilet usage represents only about 2.4% of national water usage, sewage treatment creates ecological issues that are solved by reducing toilet flow. And the Environmental Protection Agency (EPA) recently instituted a requirement that new toilets be able to flush 350 grams of waste, so most toilets should be able to handle flushing the average maximum male waste of 250 grams. That makes the dual flush the best option for both cost and water conservation. ■

RESOURCES

Aqus Gray Water System
www.watersavertech.com

Athena
www.athenacfc.com

Brondell® PerfectFlush
www.brondell.com

Toilet Rebate Finder
www.toiletrebate.com

Help your shower use less water

Green $pecs

Overall Rating

Difficulty

Green Benefits

ERIC SAYS The average U.S. household uses nearly 300 gal. of fresh drinking water every day. (In contrast, the average African household uses only 5 gal. per day.) A notable part of this usage disparity derives from the American tradition of the long, hot shower.

Although showers are the third largest consumers of indoor water use in your home (after toilets and washing machines), they add up to nearly 20% of all indoor water usage and are the largest users of hot water. By simply installing a low-flow showerhead, you can save up to 4,000 gal. of water annually, and for every gallon of hot water you can save, that's gas or electricity you don't need to use to heat it.

Beginning in 1992, federal law required all new showerheads to use no more than 2.5 gal. of water per minute (gpm). Before that, shower-heads used anywhere from 5 gpm to 8 gpm. If the average shower is 10 minutes long, upgrading your old showerhead to a low-flow model will save 25 gal. to 55 gal. of water for every shower you take, and potentially shave 30% off utility bills!

Typical Hot Water Use in a Home

Showers and washing machines use more than 60% of a home's hot water. Dishwashers, filling up the bathtub, and faucets use the rest.

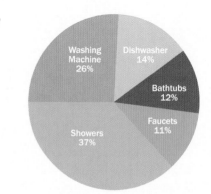

(Data adapted from U.S. Department of Energy, Energy Efficiency and Renewable Energy)

What will this project do for your home?

Every day, 3 billion gal. of water spout from showerheads in the United States, and half of that is unnecessary. Upgrading to a low-flow showerhead will immediately lower water and water-heating bills. At little cost, you can add a low-flow showerhead and cut up to 50 gal. of the 300 gal. your home uses every day.

What will this project do for the Earth?

If most old showerheads are spraying twice the amount of water needed for a good shower, we are unnecessarily wasting our most precious resource. We can save more than a billion gallons of fresh water daily simply by upgrading all of our showerheads to low-flow.

Will you need a contractor?

As long as you have a wrench, you should be able to install a low-flow showerhead on your own. You'll need the showerhead (2.5 gal. per minute or less), a plumber's wrench, pipe dope, and about 15 minutes.

Installing a low-flow showerhead Using the wrench, slowly turn the base of the old showerhead counterclockwise until it becomes loose. Don't strain or force it. Once loosened, remove the head with your hand to keep it from falling. A small amount of water will stream out of the bare pipe, so dress accordingly.

If you are renting your place, save the old showerhead in a safe place for the day you move out. Your landlord will want it back. Otherwise, just dispose of the old showerhead once your new one is installed and working properly.

Use a washcloth or paper towel to clean off the threaded end of the shower pipe and coat those threads with a layer of pipe dope. Manually twist the new showerhead onto the pipe to get it started. Use the wrench to tighten it until snug. Do not overtighten or you can damage the showerhead.

The WaterSense logo, created by the EPA in 2006, is a symbol of products that meet a high standard of water efficiency.

This ultra-low-flow showerhead features a pause button to temporarily suspend the water flow while you soap up.

What are the best sources for materials?

Your local hardware, drug, or home store will offer a line of low-flow showerheads. All showerheads sold will be no more than 2.5 gpm, but look for even lower flow amounts. Models using anywhere from 2.5 gpm down to 0.5 gpm are available.

Showerheads bearing the WaterSense® logo (at left) indicate models with the highest standard of water efficiency. The seal was developed by the EPA in 2006 to alert consumers of water-efficient appliances.

Low-flow showerheads are available in dozens of types with a variety of different features. Hand-held sprays, pulsating jets, and adjustable nozzles are easy to find, but other fun options are also available. Some manufacturers offer heads with color LED lights indicating the temperature of the water (blue is cold; red is hot). Other models have pause buttons on their showerheads to allow you to temporarily turn off the water while you soap up or shave. AquaHelix™ (www.aquahelix. net) produces one of the lowest-flow heads available at only 0.5 gpm. Most you'll find in the stores will be around 1.5 gpm to 2 gpm.

The American tradition of the long, hot shower comes from the misconception that "low flow" means "low wash." Many people assume lower pressure will leave them with soap in their eyes. However, nearly all low-flow showerheads aerate the water by pulling air into the stream, creating a feeling of higher pressure. In fact, you will probably think your new showerhead has more pressure than the old water-waster.

One small warning about switching showerheads: Because low-flow heads deliver less water, they're more likely to scald you if a toilet is flushed, due to the sudden drop in cold water pressure. Scalding shouldn't occur in bathrooms served by ample piping (3/4-in. supplies) or where thermostatic mixing valves, antiscald valves, or pressure-balancing valves have been installed. This is typically the case in newer homes, but an older home might have smaller-diameter pipes. If the temperature of your shower water currently rises when someone flushes the toilet, you should have a plumber install an antiscald valve.

How much maintenance will be required after installation?

If you see any leaking around the base of the showerhead, remove the head and add a layer of Teflon® tape to secure the connection.

If you live in an area with hard water, a showerhead can become clogged. Unscrew the showerhead and soak it for 2 minutes in a bowl of warm white vinegar to loosen any calcium buildup. That should clear the head and make it good as new.

How long will the project take to accomplish?

Installing a showerhead is easy and should take no more than 15 minutes. The real work comes from changing showering habits. To truly use less water, you have to change more than your showerhead. Shorter, faster showers will save more water and require no tools.

A shower timer is a helpful way to remind yourself not to stay too long under your new low-flow showerhead. A cute gift idea for the kids is a waterproof timer such as those sold by Ripple (www.ripple products.com). Some people use a simple sand timer, as it doesn't need batteries and lasts pretty much forever.

KEVIN SAYS I am all for the long, hot shower, so I can't say I am looking to restrict my flow. I was pleased to find out that since 1992 we have been mandated to install showerheads that flow at 2.5 gpm and that any-thing bought since that time is already meeting standards. Even the big showerheads in hotels are 2.5 gpm or less, and the fabulous steam showers use hardly any water at all. Yay for luxury. Still, there is a lot to be said for cranking it down to 1.5 gpm as long as I don't get scalded when someone flushes or freeze in a big shower space be-cause the restricted flow isn't enough to heat the area. And kudos go to all the plumbing engineers for finding ways to apply hot, steamy pressure to small amounts of water and protect sensitive skin with antiscalding valves.

What is the capital cost? $

Happily, most showerheads are fairly in-expensive. They range from around $8 to $50 for most standard fixtures. Of course, you can pay up to $300 for fancy adjust-able designer and handheld models.

Which Use Less Water— Baths or Showers?

In medieval England, it was customary for most people to bathe only once a year. When bath time came, a fami-ly would bathe in the order of the household: first Father, then Mother, then starting with the eldest children. The baby went last.

By the time baby got a chance to bathe, the water was murky from the other family members. When it came time to drain the tub, they used to caution: "Don't throw out the baby with the bath water."

You might think a bath uses less water than a shower. Depending on the size, an average tub requires 30 gal. to 50 gal. of water to fill. If you take a 5-minute shower with a 2.5 gpm showerhead, you'll use only 12.5 gal. for your shower. But if you take a 15-minute shower, you'll use almost 38 gal. Suddenly your bath might use less water.

To test the amount of water you use, put the drain plug in the bath next time you take a shower. By the end of your shower, look down to see how much the tub has filled up. That will give you a simple answer.

What financial resources are available? $

Perhaps because of the 1992 mandate, rebates for showerhead swapping are a little scarce, although they can be found in some areas. Check with your local utility, but you're pretty much on your own here.

What is the monthly cost or savings? $$$$

Showerhead savings are pretty straightforward. By switching from the mandated 2.5 gpm showerhead to one with 1.5 gpm, you can reduce shower water usage by 40%, assuming you shower for the same amount of time. That should reflect an estimated 8% savings on your water bill or $8 per month for a $100 bill. There may even be added savings to your energy bill since you'll be heating less hot water. You could shave as much as 5% to 7% off your energy bill, which could mean $10 to $20 monthly. If you have an old 5 gpm showerhead, you could knock off close to 20% of your water bill.

What is the long-term home value? $$$$$

Anyone selling a home built before 1992 would need to retrofit the showers and faucets to be compliant with the law, so you'll either pay now or pay when you sell. Newer homes will already conform to current law, but incorporating luxury fixtures with conservation in mind should make for good selling points when marketing the home. Nicely appointed bathrooms can bring big bucks, and people may pay a little extra not to feel guilty about enjoying themselves.

THE BOTTOM LINE is...

Since running the shower can account for as much as 20% of home water usage, reducing flow in the shower is a step in the right direction. It's a cost-effective way to help the planet and save money in the process. Of course, you can always boost your conservation by using a timer to reduce your shower time. The Portland Oregon Water Bureau has provided thousands of hourglass timers to its customers to help them slow the flow. Or you can always adopt the conservational approach seen on many a college campus T-shirt: "Save water; shower together!" ∎

Install
occupancy sensors

ERIC SAYS The idea is simple. When you walk into a dark room, you flip a switch and turn on the light. When you leave, you flip the switch to turn the light off. Yet, according to the U.S. Department of Energy, forgetting to turn off unneeded lights costs Americans approximately half a billion dollars a year in energy costs. This is where occupancy sensors can make a difference in the home.

Occupancy sensors are switches that turn on a light when they sense motion, and shut it off again after a set period of time. If you leave the room and forget to turn off the lights, an occupancy sensor will do it for you. When you walk back into the room, the occupancy sensor turns the light on again, giving you a reliable way of ensuring you use only the light you need.

But occupancy sensors aren't just for forgetful family members. You probably have at least one light on the outside of your home for security. Rather than leaving a porch light on all night, an occupancy sensor can turn on a light as someone approaches. The sudden flash of light could scare off any potential burglars, or provide light when you arrive home with an armload of groceries.

An estimated 10% to 15% of electricity goes toward lighting our homes. As a nation, we spend more than $37 billion annually on lighting, and one-third of that goes into our homes. By my own conservative estimate, 5% of all household lighting is wasted. Installing occupancy sensors could help reduce your electricity bill by at least that much.

What will this project do for your home?

Occupancy sensors monitor movement, automatically turning on light fixtures the moment someone enters their line of vision. Sensors, which are inexpensive and worry free, can be mounted on the wall just like a light switch or installed on the ceiling. They work well in laundry rooms,

Green $pecs

Overall Rating

Difficulty

Green Benefits

Where to Put an Occupancy Sensor

Occupancy sensors are ideal for laundry rooms, where a load of laundry in hand may prevent you from turning on or off the light. Ceiling sensors monitor the entire room, while light switch sensors look directly ahead. Both turn on the lights by sensing motion.

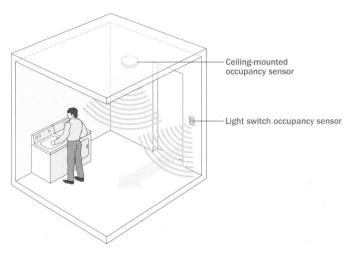

Ceiling-mounted occupancy sensor

Light switch occupancy sensor

powder rooms, the garage, over exterior doors, and in rarely accessed areas like basement hallways. In rooms where lights are frequently left on, or in rooms left unoccupied for large periods during the day or night, an occupancy sensor can cut lighting costs by nearly half. Savings will vary depending on where you live, the type of lighting, and how you use your lighting. The California Energy Commission estimates that typical savings from occupancy sensors range from 35% to 45%.

What will this project do for the Earth?

According to FlexYourPower.org, if every household in the United States installed occupancy sensors on the lights in their rarely used rooms, it would save 12 million kilowatts annually, and eliminate more than 9,000 tons of carbon dioxide, the leading cause of global warming. You'd have to plant more than 45,000 trees to remove that much CO_2 from the atmosphere!

Will you need a contractor?

Occupancy sensors are inexpensive and effective devices that can quickly and easily be installed on a wall or ceiling. You probably don't need a contractor to install a simple screw-in sensor, but you will need to be comfortable around electrical wiring if installing a permanently mounted sensor. Seek professional advice and assistance when beginning any project involving electric wiring.

DO THE MATH

Residential electricity is 31% of all electricity used.

$37 billion spent annually on electricity × 31% residential usage = $11,470,000,000 residential electricity use

If 5% of that energy is wasted, it costs households $573,500,000.
See http://www.fypower.org/images/faq/faq_enduse_chart.png for details.

What are the best sources for materials?

Your local hardware store will offer several types of basic occupancy sensor models. The most common is the screw-in type, inserted into any standard light bulb socket. The bulb is then screwed into the occupancy sensor. More advanced mounted units are hardwired into the electrical wiring. Resembling a smoke detector, these units can be mounted anywhere on the wall or ceiling. For an existing home, these will not work without opening up or fishing lines through the walls. If you're remodeling your home or putting in new wiring, the hardwired models provide a more attractive, built-in look. They work best for large, open rooms.

A third type of sensor that works for both existing and new homes is a wall switch occupancy sensor, which replaces the light switch on your wall. To install one of these, you need to be comfortable handling electrical wiring. (An electrician could do the job in less than 15 minutes.) These work best for small rooms and are ideal for infrequently used rooms, laundry rooms, and other areas, such as utility rooms, where a family member might forget to turn off the light. Living and family rooms—where people may be moving in and out throughout the day—may not be the best places for occupancy sensors for obvious reasons.

The wall switch type of sensor typically has a button to manually turn the light off and on, and models with dimmers are also available. If installing sensors in a damp location—such as a steamy bathroom—look for models specifically designed for bathroom use.

All occupancy sensors have a sensitivity dial to prevent pets or billowing curtains from setting them off. They also come with a timer to set how long the sensor should wait before turning off the lights without detecting motion. More expensive models offer dual temperature and motion sensors for more accurate readings, but basic models work just fine in the home.

Adjust the time delay (the time before the sensor switches the lights off) for as short as you can tolerate. The shorter the delay, the less the light is on and the more energy you will save. Fine-tune the settings over the first few days after you install the sensor.

How much maintenance will be required after installation?

Occupancy sensors require little maintenance but need to be unobstructed to work properly; if they cannot see, they will not turn on.

DO THE MATH

In 2007, summer demand for electricity was 783 gigawatts (GW). (By 2017, North American Electric Reliability Corporation [NERC] projects summer consumption to be 925 GW for summer and 756 GW for winter.)

783 GW = 783,000,000 kW

783,000,000 kW × 31% (amount of all electricity used to power residential homes, see p. 26) = 242,730,000 kW

5% of that energy, saved, equals 12,136,500 kW.

According to http://www.carbonify.com/carbon-calculator.htm, that 5% of energy is equivalent to 9,102 tons of CO_2 emitted into the atmosphere annually. It would take 45,511 trees to offset those emissions.

Two typical types of occupancy sensors: a wall switch sensor and a motion sensor for outdoor use.

How long will the project take to accomplish?

Screw-in sensors can be installed as easily as changing a light bulb. The mounted and wall switch models require some basic knowledge of wiring and electricity. Remember to turn off the power at your breaker box before performing any electrical work, and consult a licensed electrician before tackling any rewiring projects yourself.

KEVIN SAYS Unlike many of the energy remedies suggested in this book, in my opinion, this is one that really doesn't make a lot of sense. The main purpose of putting occupancy sensors in your home is to make up for the fact that people in your home keep forgetting to turn off the lights when they leave the room! Aside from outdoor motion sensors used for home-security lights, most occupancy sensor technology works best in a commercial setting where there is either high traffic or no traffic. For myself, I couldn't find many places in the house where lights were left on willy-nilly. I thought about screw-in adapters, but none of my fixtures would accommodate the bulky sensors.

Ultimately, I did the math and selected CFLs as an alternative, which, by the way, apparently don't really work well with occupancy sensors since CFLs need to be on for 20 minutes or more for maximum efficiency and the sensors will cause them to go on and off in short cycles.

What is the capital cost $

Occupancy sensors for the home range from $15 to $50 per unit. Since most homes won't have use for more than two or three, you won't have to invest a lot initially.

What financial resources are available?

Most cities and public utilities offer plenty of incentives for commercial application of occupancy sensors, but few if any exist for the homeowner.

What is the monthly cost or savings? $

Green sources argue that occupancy sensors can reduce lighting costs by up to 50% in rooms where lights are frequently left on. The sources estimate 30% to 80% savings in corridors and 45% to 80% in storage areas; however, these estimates are all based on commercial use. There is no monthly cost for sensors unless they turn on lights when you ordinarily wouldn't turn them on yourself. Since you probably won't need more than a couple of these in your home, even a 50% usage savings on those lights would amount to less than a dollar a month, meaning it would take you more than a year to recoup your investment.

What is the long-term home value?

The cost of sensors is so minimal and the usage is generally so isolated in a typical home that it won't have any real value. Not only that, but it could be detrimental if you one day decide to sell your home, and the sensors are located in inappropriate places—the disco effect may make the buyer think your electrical system is having problems.

THE BOTTOM LINE is...

This is an idea that sounds good, but isn't great in practice. Truth be told, since most residential applications leave you having to make a practical choice between CFLs and sensors, you would save more energy and money by simply replacing incandescent light bulbs with CFLs since they bring a 75% savings per bulb versus an optimistic 50% with the sensor. ■

RESOURCES

Occupancy sensors can be found at your local hardware store or online. Here are a few suggested manufacturers:

Evecto
www.evecto.com

Greengate
http://greengate.
coopercontrol.com/
coopercontrols

Leviton
www.leviton.com

Lutron®
www.lutron.com

Sensor Switch
www.sensorswitch.com

WattStopper
www.wattstopper.com

Install smart strips and kill switches

Green $pecs

Overall Rating

Difficulty

Green Benefits

ERIC SAYS Our homes are all too good at doing their job of providing energy when we need it. Flip any switch, and the power comes on immediately. The appliances and equipment in our homes are overachievers at their jobs, too. They sit in standby, always ready to come to life at a moment's notice. If something is plugged into the wall, even if it's not on, it draws electricity. From kitchen appliances, cell phone chargers, and power adapters with no "off" switch to appliances with clocks and the living room television, our homes consume electricity even when no one is home. We call this demand of energy "phantom loads" or, more appropriately, "vampire loads," since they suck energy. Vampire loads lurk all over the home, and while the amount of power used by each device is relatively small, their sum can add up to more than 10% of your electricity bill.

There are several simple ways to slay vampire loads:

- **Unplug:** Some people believe equipment lasts longer if it is never turned off. This incorrect assumption carries over from the old days of big mainframe computers. The reality is that shutting down and unplugging devices not only guarantees that you'll save energy but also saves on the devices' wear and tear. Unfortunately, this is not always convenient or easy, especially when things are plugged in behind the sofa.

- **Power strips:** A power strip allows you to plug in several items that are controlled by a single switch. Installing basic surge protector power strips allows you to flip the switch off when not needed. Depending on its location, a power strip might be a little easier to reach than several plugs in different locations.

- **Smart strips:** "Smart" power strips are available that sense when power is being drawn and shut off automatically. As simple to

Any device with a standby indicator light is a notorious source of vampire loads. Standby mode wastes energy by keeping a device in a state of constant readiness.

install as a regular strip, you don't need to worry about vampire loads ever again.

- **Kill switch:** If you're building or remodeling a home, you can install a standard light switch to shut off certain outlets in a room. This makes the power source easy to reach and makes it obvious to see if something has been left on.

A "smart" power strip not only tells you how much energy you're using but also automatically switches off when not in use.

What will this project do for your home?

The amount of the vampire load varies depending on the type of appliance. An electric toothbrush charges only an hour a day and will cost around $1 a year in vampire load. A desktop computer might only be used after work, but will still cost $34 per year. While these are relatively small amounts, a widescreen plasma television, however, uses an active standby mode that can cost up to $160 a year to continually warm the plasma and keep the television at the ready. In the average home, 75% of the electricity used to power home electronics—cable boxes, DVD players, video games, stereos—is consumed while the products are turned off. That's money that could stay in your pocket. As an added benefit, when appliances are unplugged, less heat is created in your home. This lowers the need for air-conditioning in the summer and potentially saves even more.

What will this project do for the Earth?

Eliminating vampire loads is one of the easiest and least disruptive things you can do to conserve energy in your home. Individual phantom loads might be fairly small, but appliances are plugged in around the clock and can account for up to 22% of their energy use. In total, vampire loads can consume 5% to 10% of all residential electricity. If saved, that is enough energy to power more than 10 million homes in the United States for as long as the vampire loads are saved (an hour of saved vampires will power those homes for an hour, etc.).

Vampire loads cost homeowners more than $3 billion in energy each year. That is equivalent to the output of 18 power stations. Preventing such energy waste will go a long way to slowing global warming and protecting our environment.

DO THE MATH

According to the Department of Energy, national residential electricity consumption in 2004 was 1.29 billion megawatt hours (MWh), 5% of which is 64 million MWh.

That wasted energy is equivalent to the output of 18 typical power stations.

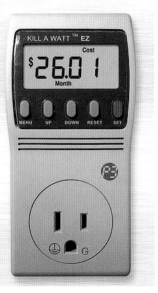

An inexpensive energy meter can calculate how much energy a device uses when it's on or in standby mode.

Will you need a contractor?

Installing a kill switch to power outlets requires an electrician and is best done on new construction or during a remodel of your home. Power strips and smart strips can be installed by anyone in just a couple of minutes. If using standard power strips, locate them in a spot you can easily reach to turn them on and off.

What are the best sources for materials?

Your local hardware store, drugstore, or convenience store will carry a standard power strip. Office-supply stores offer strips with higher surge protection, important in the event of lightning strikes and power surges, and online retailers such as Amazon.com® (www.amazon.com) and Gaiam® (www.gaiam.com) offer a variety of models.

Smart strips sense when a device is on and drawing power and will turn it off automatically. They cost a little more than standard power strips. Some models indicate the amount of energy each plug is using, but a basic smart strip works just as well. Terra Pass (www.terrapass. com), WattStopper (www.wattstopper.com), and Smart Strip (www. smartstrip.net) all carry a variety of reliable smart power strips and can be ordered online.

If you're curious about the exact size of a particular vampire load, an energy meter can show you. Watts Up (www.wattsupmeters.com) and P3 International (www.p3international.com) offer simple, low-cost meters. Simply plug an appliance into the meter and it indicates the amount of energy needed to use the device and its potential cost. These meters don't save energy; they simply show you the size of the vampire load.

If you are already mindful of your energy use or don't have many standby devices, you probably don't need to install smart strips for everything in your house. Look for devices that are warm, even when not in use. Devices with a standby indicator, ready light, or clock are prime suspects. If you already have a clock in your kitchen, unplug other countertop appliances (coffee maker, blender, microwave). Plug them in when needed and don't worry about the clock not being set.

Some devices need to be plugged in all of the time, despite vampire load. Certain cable boxes, satellite receivers, and digital video recorders (like TiVo™) need to be on to download and record programming. For these, there is little you can do to reduce their consumption.

How much maintenance will be required after installation?

Once installed, power strips require virtually no maintenance; just keep them dusted and dry. Check all of the electrical cords around your home. Frayed cords are a fire risk and should be replaced. Power strips often have a fuse or breaker, so check them regularly.

How long will the project take to accomplish?

Take a walk through your home and make a list of the appliances that can be unplugged or switched to a smart strip. Since most strips can accept four to eight plugs, you'll probably need only three or four strips total. Installation can be completed in less than an hour. By the way, if you're pulling out furniture to reach the outlets in the wall, it may be a great opportunity to vacuum.

KEVIN SAYS Obviously, vampire loads present a monster household problem. (I couldn't resist.) Still, some loads are beyond our control and others provide significant convenience. Many appliances like ovens and refrigerators draw energy, and we can't and shouldn't shut them off. Other appliances—the microwave, for example —have clocks that draw a constant trickle of power. I, for one, can live with that minor vampire load because I don't want to reset the clock every time I use it, nor do I want it flashing at me. I will happily pay the extra estimated $1.38 per year not to deal with it.

As a proud owner of an LCD television, I was happy to discover it was a winner over the plasma TV as far as pulling a lower load in the off stage. I don't mind putting a switch on my TV, but my DVR/cable box has to stay plugged in so I don't miss downloading my reality TV. I think the smart thing here is to use an unswitched plug for the DVR or plug it in separately and let the TV go to sleep for the night by disconnecting the power.

What is the capital cost? $

Solutions for vampire loads can be simple and cheap. Unplugging appliances when you are done using them is easy and free. There are $5 individual kill switches that go on the wall socket, which seem sort of

RESOURCES

Gaiam
www.gaiam.com

Greenswitch
www.greenswitch.tv

P3 International
www.p3international.com

Smart Strip
www.smartstrip.net

Terra Pass
www.terrapass.com

Watts Up
www.wattsupmeters.com

WattStopper
www.wattstopper.com

silly since you have to bend down to the switch anyway. Smart strips make some sense since they cost $15 to $30 and do all the work for you once you install them. A typical house may need three or four strips to cover the computer and TV areas. Even the largest house should be able to resolve their issues for less than $150.

What financial resources are available?

Other than the occasional coupon offering for the big-box stores, you probably won't find much in the way of savings for reducing vampire loads.

What is the monthly cost or savings? $

Most experts put typical vampire loads at 5% to 7% of the home power draw. Videophiles with multiple big screens will be on the higher side of that range. Since it is unlikely you will be able to remove these loads entirely, a reasonable savings expectation might be 4% to 5% of your monthly energy cost. Figure $8 to $10 on a $200 monthly bill. It's not huge money, but you will pay yourself back for the smart strips in the first year or two.

What is the long-term home value?

The solutions for vampire loads are all add-on solutions, so they really have zero impact on the value of your home. The only possible exception here could be built-in appliances that draw load even when they are shut off. Currently, a buyer would not take this into consideration, but as technology improves and awareness increases, smarter appliances will become more widely available and you may be looking at the need to upgrade.

THE BOTTOM LINE is...

The real issue here is prioritizing according to what appliances you really need plugged in all the time and those you don't. You may want to keep the automatic coffeemaker ready to brew in the morning and keep the big-screen TV and DVR plugged in, but unplug the computer and replace old power strips with new smart strips. You'll save a little money *and* feel good about saving the planet. ▓

Install a programmable thermostat

ERIC SAYS The old-fashioned thermostat is a fairly dumb device. It switches on a heater when the temperature drops below a set degree and turns the heater off once the temperature rises above that setting. So the heat comes on no matter what time of day it is, likely wasting energy while you sleep or when you're not even home. These models are also inaccurate, using an old-fashioned manual dial, and are often located on a drafty exterior wall where cold air leaks in from the outdoors and causes more inaccuracies.

A programmable thermostat solves these issues. It is a smarter version of the original because it operates only during the times you set. For example, a programmable thermostat could lower the heat from 68°F to 62°F at 10 p.m. every night, when you're bundled under the covers in bed. It could also be programmed to return the room to a more comfortable temperature 30 minutes before you wake up. A programmable thermostat can also turn off the heat while you're at work or school during the week, and turn it back on just before you arrive home so you have a toasty living room waiting for you. You get the convenience of setting your home's temperature according to your preferences and can capitalize on savings by not wasting energy when you don't need heating or air-conditioning.

What will this project do for your home?

The average household spends more than $2,000 a year on energy bills—nearly half of which goes to heating and cooling. Expect to save around 15% a year for heating and cooling by setting the thermostat back just 5 to 10 degrees while you're at work or school. Once set correctly, a programmable thermostat can cut your heating and cooling bills by 20% to 30% annually.

Green $pecs

Overall Rating

Difficulty

Green Benefits

A programmable thermostat can control the heating and cooling in a home based on set schedules throughout the day.

Look for programmable thermostats bearing the Energy Star logo. They come preprogrammed with a range of settings (hourly/daily/weekly).

What will this project do for the Earth?

Heating, cooling, and lighting our homes belches out one-third of our nation's total carbon emissions and sucks hundreds of dollars out of our wallets in the process. Even a 10% reduction in every home would amount to a critical reduction in greenhouse gas emissions.

Will you need a contractor?

The installation of most programmable thermostats is relatively safe and simple because they are connected to low-voltage wiring. Replacing your old thermostat with a new programmable model requires the removal of a few screws and twisting a couple of wires together. But dealing with wires is always risky, so be sure you know what you're doing or consult an electrician. Always turn off the main power to your home before performing any electrical work. And be aware that only qualified electricians or heating, ventilation, and air-conditioning (HVAC) contractors should install programmable thermostats for electric baseboard heaters because the wiring carries higher voltages.

What are the best sources for materials?

Your local hardware store should offer at least one model of programmable thermostat, but you'll find a greater selection online at places like Amazon.com (www.amazon.com), Lowe's® (www.lowes.com), and The Home Depot (www.homedepot.com). Energy Star–qualified programmable thermostats come preprogrammed with settings that are intended to deliver savings without sacrificing comfort. Look for the Energy Star logo on the packaging.

The location of your thermostat can affect its performance and efficiency. If upgrading your existing thermostat, you will be stuck with the current location. If remodeling or building a home, choose a nondrafty location away from vents and windows. Also avoid placing thermostats in direct sunlight. Drafts and sunshine can give a thermostat an inaccurate temperature reading. If the thermostat is located on an outside wall, simply place a foam thermostat gasket under the wall plate. It costs a few cents and will prevent outside air from affecting the temperature.

Programmable thermostats typically offer a number of programming options: daily programming to run the same schedule each day; weekday/weekend settings; and seven-day programming, which allows for a different schedule every day.

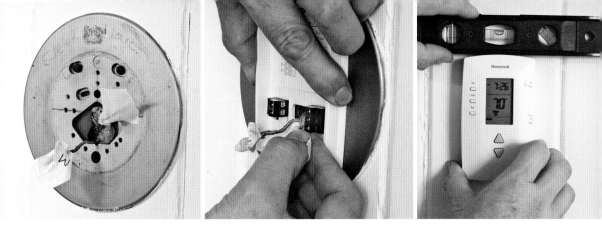

Use masking tape to label the electrical wires, noting where they attached to the old thermostat.

Connect the wires to the new thermostat according to the manufacturer's instructions.

Mount the unit to the wall plate and use a level to ensure it sits straight.

Installing a programmable thermostat You'll need a screwdriver or drill, electrical wire caps (which often come with the thermostat), a level, and masking tape (to hold the loose wires). Before starting installation, read the instructions that came with your programmable thermostat as manufacturers' instructions will differ slightly from model to model.

1   Turn off the power supply to the thermostat by flipping off the appropriate circuit breaker. (If you don't know what a circuit breaker is, stop work and call an electrician!)

2   Remove the old thermostat, including the wall plate. Older thermostats used mercury-filled glass tubes. Always use caution when handling mercury. Check with your local recycling or waste company for instructions on recycling the mercury; never throw it into the trash.

3   Most new thermostats will have either two or four wires (two for one unit; four for separate heating and cooling). Using masking tape, label the wires and note where they were attached to the old thermostat. Tape them lightly to the wall to prevent them from falling into the wall cavity. Trust me—you only make that mistake once.

4   Follow the instructions included with your programmable thermostat to mount the new wall base to the wall. Position it to cover the area where the old unit used to be.

5   Carefully connect the wires according to the manufacturer's instructions. Install batteries, as directed.

6   Mount the unit to the wall plate. Use a level to make sure that it isn't crooked.

7   Turn the power back on. Test the thermostat to make sure it works properly.

8 Program your new thermostat following the manufacturer's instructions. Most programmable thermostats are digital. They can be hard to program (similar to a VCR), so keep the instructions handy.

Electric or radiant heating systems may require different types of programmable thermostats. When choosing a programmable thermostat, make sure that the thermostat is compatible with your home's HVAC system. Programmable thermostats are generally not recommended for heat pumps since they can cause the pump to run inefficiently. But specially designed programmable thermostats for heat pumps are available.

How much maintenance will be required after installation?

Programmable thermostats don't need much maintenance (some thermostats contain a battery in case of a power outage that will need to be replaced every year or so), so long as the settings are working for you. Look for regular periods of four hours or more when you can lower the temperature settings. Periods when no one is home and during sleep hours are the most common times people adjust the temperature. When needed, you can manually override a thermostat's settings to make it warmer or cooler without altering the programmed settings. If you consistently override the settings, change the program times to better reflect your preferences. Most units also offer a vacation setting for when you're planning to be away for an extended period of time.

Most homes have only one thermostat, and only one "zone." In a single-zone home, it isn't possible to vary the temperature by room. The heat is either on or off everywhere. If remodeling or building a home, consider having different zones installed to allow you more control. The living areas could be in their own control zone, the bedrooms in another. Infrequently used areas, such as guest rooms or basements, could have their own control zone. Each zone can be controlled by a programmable thermostat.

Higher, Not Faster

It is a myth that a furnace has to work harder to warm a space back to a comfortable temperature after the thermostat has been lowered. This isn't true and is not how home heating works. The energy required to reheat your home is the same as the energy you save by letting your home cool down. The real energy savings come from the furnace being off or running at a lower temperature for several hours. The longer the programmable thermostat holds the temperature at the lower level, the more energy you save.

Cranking up the heat to 90°F or down to 40°F will also not heat or cool your house any faster. The air coming out of the vents is still the same temperature. Turning the heat way up only means the thermostat will not turn off until it reaches that temperature. Adjust the thermostat to the final temperature you want; setting it very high or very low will not speed up the process.

How long will the project take to accomplish?

Installing a programmable thermostat can be done in less than an hour. The hardest part is programming the darn thing.

KEVIN SAYS I really don't know how most people live with a set temperature in their homes. My wife and I have such different body temperatures I wonder sometimes if we are of the same species. We constantly adjust the thermostat in our house, especially when we know the other one isn't looking. Still, we live in an old house that is hard to regulate compared with modern homes.

However, if you can agree upon a set temperature for the various seasons, then a programmable thermostat should result in savings for you. Big houses in temperate climates apparently can save even more, and it makes sense to program zones in areas of the home that are used less, like bedrooms.

It seems that the bugs have been worked out on programmable thermostats for floorboard heaters and radiant heat. Amazon.com has highly rated products that are very affordable.

The biggest problem of switching out an old thermostat is if it is a mercury-filled unit that needs to be recycled. The EPA lists recycling sources for mercury by state (www.epa.gov/wastes/hazard/tsd/mercury/collect.htm). If you can't find something there, you can call a local HVAC repairman who will probably let you drop it off.

What is the capital cost? $

Programmable thermostats are reasonably priced regardless of whether they are for forced air, radiant heat, or floorboard heat. They start at around $30 and can go as high as $150; most homes will only need one or two units.

What financial resources are available? $$$

It seems that plenty of cities want to help you dump the old thermostat. Cities like Charlottesville, Virginia, will give you up to $100 to cover the cost and installation of a new programmable thermostat, making it

RESOURCES

EPA Directory of Mercury Recycling Sources
www.epa.gov/wastes/
hazard/tsd/mercury/
collect.htm

virtually free. There is no central location to find these, so check with your city or local utility for available rebates.

What is the monthly cost or savings? $$

Savings estimates for programmable thermostats are literally all over the map. Mild climates see greater savings than volatile climates because there is longer usage of heating and cooling systems to compensate for extreme temperatures in the latter. Several web tools and sites claim savings of $180 annually, but the math doesn't quite add up.

The government estimates that the average heating and cooling cost in a household is around $1,000 per year. You can save roughly 15% of that by simply adjusting your thermostat up in the summer or down in the winter for eight-hour periods. A programmable thermostat allows you to do that conveniently. With a fancy model you might find other times to program the temperature, saving a little more. But even with 15% savings, you reap only $150. But still, $150 will more than cover the cost of a thermostat in the first year.

What is the long-term home value? $

You won't reduce the time your house is on market with one of these things, but you will save enough to get the purchase price back. Of course, a new thermostat won't fix an old, clunky heating system, but at least you can tell prospective buyers it's one less thing they have to buy.

THE BOTTOM LINE is...

This is a good idea all around. Thermostats are often forgotten unless we feel overly cold or hot, so often we waste energy by not making moderate adjustments. Installing a programmable thermostat saves money, and you might even get reimbursed for the upfront cost. Sure, you could adjust the temperature manually, but it's nice, as TV infomercial guy Ron Popeil says, to "just set it and forget it!" ▪

Insulate your water heater

ERIC SAYS If your home is like most, hot water is produced in a hot water heater. This large tank usually sits in a garage, closet, or basement and slowly heats up a vat of water. Think of it as a 60-gal. thermos with a built-in heater. If the stored water drops below a certain temperature, the heater automatically heats it back up again, so you always have hot water ready to go at any moment. Providing this endless supply requires energy—and lots of it. Nearly 20% of all of the energy used in the home goes just to the water heater, making it the second-largest energy user in homes after heating and cooling.

Whether your hot water heater runs on gas (a fossil fuel) or on electricity (created by burning fossil fuels), it contributes to global warming and to our ongoing energy crisis. By reducing the temperature of the heater—with a timer, thermostat, or simply by turning it down—you can reduce its energy consumption. And just as you insulate the walls of your home to hold in heated or cooled air, insulating your hot water heater is equally important. The time it takes to install water heater insulation will be well worth the payback in energy savings.

What will this project do for your home?

Insulating a water heater tank reduces the heat losses or standby losses from keeping the water sitting around waiting for you to turn on the hot water faucet by 25% to 45%. This translates into as much as a 9% savings in total energy usage.

If you have an electric water heater, you can save an additional 5% to 12% of your energy usage by installing a timer that turns the heater off at night or while you are at work and don't need hot water. Plus, a well-insulated water heater means there will be less chance of your running out of hot water during a shower, since the heat remains in the tank.

Green $pecs

Overall Rating

Difficulty

Green Benefits

We use 122 billion Kilowatt hours (kWh) of electricity each year; 9% of that electricity could be saved by insulating a water heater, equaling 10.98 billion kWh.

In 2006, the average annual residential electricity consumption was 11,040 kWh.

10.98 billion kWh
/11,040 kWh per year
= 994,565 homes powered

See http://www.eia.doe.gov/emeu/recs/recs2005/c&e/detailed_tables2005c&e.html and http://www.eia.doe.gov/emeu/recs/recs2005/c&e/waterheating/pdf/tablewh3.pdf for additional info.

What will this project do for the Earth?

In the United States alone, water heating is responsible for 122 billion kilowatt hours (kWh) of energy used annually. If everyone insulated their hot water heaters, nearly 11 billion kWh of that energy would be saved—enough to power one million homes for a year.

Will you need a contractor?

No. Insulating your hot water heater tank is simple, and quick, inexpensive and can be done in about an hour.

What are the best sources for materials?

Precut insulation kits are widely available at your local home center stores. Before you head out to the store, make note of the gallon capacity of your hot water heater. Be sure to choose a wrap with an insulating value of at least R-8; higher numbers are better. Read the manufacturer's instructions carefully before wrapping the heater, and be sure not to block any vents or get too close to the pilot light (gas flame).

For this project, you'll need a water heater insulation kit, tape measure, duct tape, and a utility knife.

Wrapping a water heater

1 Measure and cut the insulation for the top of the tank to fit around the piping coming out of it. Cut it a few inches larger than the top so you can tape the insulation to the tank sides. Do not insulate the top of a gas water heater.

2 Fold the edges of the top insulation down and tape to the sides of the tank.

3 Position the insulating blanket around the tank. Be sure the blanket ends do not overlap on top of the access panel on the side of the tank.

4 With the insulating blanket wrap in place, tape it (or tie it, depending on what is included in the kit) around the tank. Evenly space the tape or ties, making sure they do not cover the access panel. The tape or belts should be snug but should not squeeze the blanket more than 20% of its thickness.

5 Fit or cut the blanket to go around the pressure-relief valve and the overflow pipe sticking out of the side of the tank. Keep the blanket away from the drain at the bottom and the flue at the top.

6 By pressing on the blanket wrap, mark the corners of the tank access panel. Use a utility knife to make an X-shaped cut in the

insulating blanket from corner to corner. Fold these triangular flaps underneath the insulating blanket.

The installation of insulating blankets on gas water heater tanks is more complicated than for electric models. Follow the same steps as above, but make sure the airflow to the burner isn't blocked or obstructed and that the thermostat is left uncovered by the wrap. Never install an insulating blanket on a leaking tank. If your tank is leaking, you may need a new water heater.

The fastest and easiest way to cut your hot water bills is by simply turning down the temperature on your water heater. Most manufacturers preset the thermostat to 140°F, but turning it down to 123°F is ideal. That temperature is hot enough to kill bacteria but not hot enough to scald anyone.

Reducing your water temperature to 123°F also slows mineral buildup and corrosion in pipes, helping a water heater last longer. For each 10°F reduction in water temperature, you can save from 3% to 5% in energy usage.

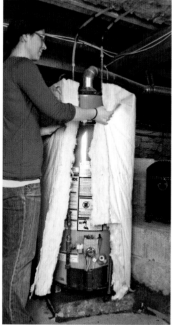

Wrap the insulation blanket around the water heater, avoiding the flue at the top and the drain at the base.

Secure the blanket in place with ties or tape.

Make sure the blanket is not gaping, bunched, or torn, which will reduce the effectiveness of the wrap.

Improve Water Heater Efficiency

If you have a tank water heater, here are some tips to cut down the energy consumption:

1 **Turn down the heat.** Water heaters have a simple thermostat to set the water temperature. If the temperature is set too hot, then you are mixing it with cold water at the tap to cool it down. This wastes energy. If you don't need the water that hot, turn down the thermostat.

2 **Insulate your hot water pipes.** Your local hardware store carries insulation for hot water pipes. These $10 sleeves install in seconds and increase your water temperature by 2°F to 4°F. See chapter 8, p. 46, for more information.

3 **Replace older heaters.** If your heater is older than 10 years, it may be time to replace it with a new, more efficient model. Newer electric heaters also have timers that help save energy by heating water according to the schedule you set.

4 **Fix leaky faucets.** A leaking water heater is a sign of future trouble and is costing you money in the meantime. A heater that drips once per second costs approximately $4 a month in lost energy and even more in water costs.

5 **Insulate the room.** Water heaters are often put in cold locations, such as basements or garages, which causes the heater to work much harder than it needs to. If possible, insulate the room where the water heater is kept.

6 **Install a drain recovery system.** As hot, soapy water goes down the shower drain, a heat-recovery drain steals the heat from this water and circulates it back to the water heater. The $400 to $600 drain can save $200 to $350 a year in hot water costs.

7 **Turn it off when you're out of town.** If you are going away on vacation, turn off the water heater. Electric water heaters can be shut down at the breaker box; gas heaters can often simply have the thermostat turned down low. That way you will not need to relight a gas pilot light when you return.

How much maintenance will be required after installation?

Water heaters can last 10 to 15 years, and periodic water heater maintenance can significantly extend your water heater's life and minimize loss of efficiency. Read your owner's manual for specific maintenance recommendations.

How long will the project take to accomplish?

Insulating your hot water heater should take no more than an hour, less if you have a helper.

KEVIN SAYS Making small changes to your water heater is a simple and effective way to save energy and money. As a fan of hot showers, I was relieved to learn that I can turn down the setting and kill bacteria at a lower temperature that still makes my skin tingle. Most of the remedies like putting on an insulation blanket and turning down the thermostat are fairly simple. The timer is more work and you really need to be sure that a late-night shower is out of the question or you may be in for a chilly shock at 3 a.m.

What is the capital cost? $

Obviously, turning the thermostat down is free and insulating blankets are inexpensive. No more than $25 should get you a top-of-the-line insulating blanket that you can install in an hour. If you decide to go the timer route, you'll pay from $60 to $100 plus installation if you need an electrician, which might cost you another couple hundred dollars.

What financial resources are available? $$$

A number of cities and states offer rebates of $5 to $10 for blankets. A few water districts occasionally offer free blankets, so it's worth calling your local utility to find out what's available. Timers don't stimulate the same rebate activity.

What is the monthly cost or savings? $$$$

Hot water usage varies from home to home, but if you are saving 25% to 45% of your water heater energy—which represents roughly 20% of your energy bill—you can count on a 5% to 9% savings (or $10 to $18) on a $200 monthly bill. That means the blanket pays for itself in a couple of months, and it's pure savings after that.

What is the long-term home value? $

Sadly, you likely won't get more for your house because of a wrapped heater, but you get a lower energy bill, which may help in the process.

THE BOTTOM LINE IS...

This remedy is a hot one for sure! Who wouldn't want to make a small investment of time and money to save more than $100 annually? The timer may not be perfect for everyone, but turning down the thermostat and adding an insulating blanket are smart things every home-owner should do. No need to ponder more—it's a wrap! ■

Insulate your pipes

Green $pecs

Overall Rating

Difficulty

Green Benefits

ERIC SAYS Insulating your water pipes is one of the easiest things to do to cut the energy consumption of a water heater. It helps hold the heat in and reduces the work the hot water heater needs to do to keep your water warm. Heating water consumes nearly 20% of the energy used in your home, making it the third largest source of residential energy consumption (after heating/cooling and appliances). But wait, there's more. Insulating your pipes also delivers hot water to your faucets and showerheads faster, thereby saving water and energy.

However, it isn't just your hot water pipes that need insulation. By also insulating cold water pipes, you prevent them from dripping condensation and absorbing the air temperature around them. In fact, you should insulate any easily accessible and visible pipes in your home. You won't be able to reach all of them, but the more insulation you can add, the more savings you'll see on your energy bill.

What will this project do for your home?

It makes sense to help keep your water hot on its way to the shower. Insulating hot water pipes raises the effective hot water temperature by 2°F to 4°F. That means you can turn down the thermostat on your hot water heater by 2°F to 4°F, saving you around 5% on your hot water bill.

What will this project do for the Earth?

The energy saved by insulating water pipes reduces the average household carbon dioxide emissions by 52 lb. per year. That is 15 times more effective than giving up plastic bags for a year. And every 10% reduction in water temperature results in a 3% to 5% reduction in energy costs.

Will you need a contractor?

Although a contractor or handyman might be helpful, this is definitely a project you can do yourself in an afternoon. But be careful when touching the hot water pipes as they could be holding water that is 120°F or more.

What are the best sources for materials?

Your local hardware store will carry everything you need. You can even order materials online to be delivered to your door from sites such as The Home Depot (www.homedepot.com), Lowe's (www.lowes.com), and Ace Hardware (www.acehardware.com).

There are two types of pipe insulation commonly available: fiberglass pipe wrap and foam pipe sleeves. Similar to the fiberglass you might see in your attic, fiberglass insulation wraps around the pipe and is held in place with tape. While it is slightly cheaper than the foam sleeves, it is much harder to install. You have to wrap the fiberglass around each pipe manually and secure it. The wrap comes in rolls 3 in. wide and up to 35 ft. long. You need gloves, goggles, and a mask to install any fiberglass material.

I prefer using foam pipe sleeves. Resembling long pipes, the sleeves slip over a pipe in seconds. Simple to install, this insulation typically comes in 6-ft. lengths and can then be cut to fit the length of your pipe. The sleeves are held in place by adhesive, tape, or cable ties. Some fancy versions of the sleeves are available with adhesive edges to glue the foam tube together after installation.

Insulating pipes with foam sleeves Using a tape measure, walk around your garage, basement, and/or crawlspace, and measure the length of the pipes you will be able to insulate. Also note the diameter of the pipes so you can select the proper size insulation.

1 Remove any dirt and grease from the pipes by cleaning them with a rag. Make sure the pipes are dry before installing the insulation.
2 Select a piece of foam insulation that will wrap around the pipe completely. Hold a length of it along the pipe and cut it to length with a utility knife, cutting the ends at an angle where pipes turn. Try to make the closest fit possible.
3 Slip the insulation onto the pipe, and secure the insulation every 12 in. to 18 in. with tape or a cable tie. Cable ties install quickly and are easy to remove. Tape holds the sleeves better, but takes longer to install. Try to avoid using duct tape as it will eventually peel away from the foam.

DO THE MATH

A typical water heater holds 80 gal. of water and water weighs 8 lb. per gal., so the heater holds 640 lb. of water.

A BTU is the amount of energy needed to raise 1 lb. of water 1°F.

If the water coming into a home is 55°F, it needs to be raised 68°F or more to reach 123°F.

640 lb. of water × 68°F = 43,520 BTU needed for water heating

By insulating your pipes, you can lower the temperature on your hot water heater by 4°F and still have the same effective hot water temperature, all while saving more than 5% on your bill:

640 lb. of water × 4°F = 2,560 BTU

2,560 BTU / 43,520 = 5.88% savings

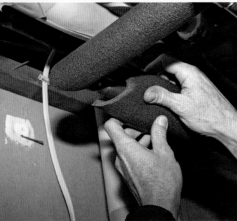

Using foam pipe sleeves ensures a snug fit around pipes. Cut the sleeve to size, and snake it around corners by angling the edges of the foam.

4   Be careful to cover the entire pipe when wrapping around corners and bends, or the insulation loses its efficacy. Cut or miter the insulation as needed to turn the corners. Follow the manufacturer's directions in the insulation packaging.

Try to insulate every water pipe that is visible and accessible. At a minimum, insulate the first 5 ft. to 10 ft. of the hot water pipe coming from the water heater and the first 3 ft. of the cold water inlet pipes coming into your home. When installing pipe insulation near an oil- or gas-fired water heater, keep the insulation at least 6 in. away from the flue.

How much maintenance will be required after installation?

With proper installation, pipe insulation should be maintenance free. Periodically check along the length of the pipes for water leaks or gaps in the insulation.

How long will the project take to accomplish?

Armed with the proper-sized sleeves and tools, you should be able to insulate your pipes in a single afternoon, depending on the size of your house.

KEVIN SAYS I can't find a big argument against insulating pipes. However, this project really only applies to situations where pipes are accessible. Houses built on a slab won't work since the pipes are embedded in the concrete, and you may be deterred if your crawlspace is tiny. For houses with exposed pipes in basements, you will have to spend a bit of time and money, but you will get decent payoff.

The thought of doing any work with that fiberglass insulation stuff makes my arms itch. Luckily, foam is reasonably priced and a lot easier to work with. The best part is that if you have extra, you can slice it in half and let your kids make roller coasters for marbles or other craft projects with it.

What is the capital cost?　$

The size of your house is the biggest factor in determining the total cost of pipe insulation, but even insulating pipes in a big house is cost-effective. Foam insulation costs about 35% more than fiberglass, but it's a better insulator and easier to install. Figure a typical house might need up to 50 ft. of insulation at most. With 6-ft. pieces of insulation ranging from $1.50 to $6 for self-adhesive foam, materials will cost from $13.50 to $54. You can stop there if you do it yourself or add a few hundred more to get a day laborer to install it for you. The most you are likely to spend is $350 total.

What financial resources are available?　$

Not much available here. Check with public utilities for general tax breaks or incentives, but most credits go to commercial applications.

What is the monthly cost or savings?　$

Insulating hot water pipes increases the heat about 2°F to 4°F, which means you will save roughly 5% of your water-heating energy. You'll also save on the cost of water (if you pay for it) since you won't have to wait as long for hot water to get to the faucet. All in all, you might save $30 in a good year. Not much to get excited about. If you install it yourself, you'll recoup the cost in less than two years.

What is the long-term home value?　$

There aren't really enough dollars at stake to worry about recouping the cost at sale. However, this retrofit makes for an impressive showing to prospective buyers about the house being energy efficient and might get you a few hundred extra in their initial offer.

THE BOTTOM LINE is...

There's no reason not to do this. At the very least, wrap the lead pipes for both hot and cold. Foam insulation is easier to install and worth the extra few bucks. Do it yourself. You'll impress your friends that you did something handy and planet worthy. ■

Seal your ducting

Green $pecs

Overall Rating

Difficulty

Green Benefits

ERIC SAYS Leaking ducts in the heating and cooling system of a home can waste hundreds of dollars a year. Ducts leak because of the way they are constructed. All ducts are assembled from shorter lengths, and each connection is a potential source for air leakage because ducts expand and shrink as the temperature of the air within the duct changes. Air can leak both in and out of ducts, forcing air conditioners to work harder when hot air seeps in and causing furnaces to stay on longer if cold air works its way into the system. Duct leaks also make it difficult to keep the temperature in a home comfortable. Stuffy rooms and chilly spaces are signs that there may be leaks in your ductwork.

Even if you live in a new home, don't assume your ducts are sealed. Although ductwork contractors know that sealing is a "best practice," ducts are typically only sealed at the architect's or contractor's request or if the home is built to a green standard, such as GreenPoint Rated (www.builditgreen.org), Built Green® (www.builtgreen.org), or LEED for Homes (www.usgbc.org/LEED/homes/). Most other homes probably do not have sealed ducts.

What will this project do for your home?

A typical home contains about 180 linear feet (lin. ft.) of ductwork. Nearly 30% of a home's conditioned air leaks out of these ducts, wasting energy in the process. Sealing ducts saves you a significant amount of energy and money and also improves the indoor air quality in a home—ducts accumulate more than 40 lb. of household dust each year. Leaky ducts spread that dusty air throughout the home, while sealed ducts deliver cleaner air.

Duct Leaks: Where and Why

This diagram shows where air can leak in and out of a home's ducts and the common problems that lead to the leaks. Your home may have poorly performing ducts if you have high utility bills in the summer and winter, if you have stuffy rooms that never seem to feel comfortable, or if there are rooms in the house that are difficult to heat and cool.

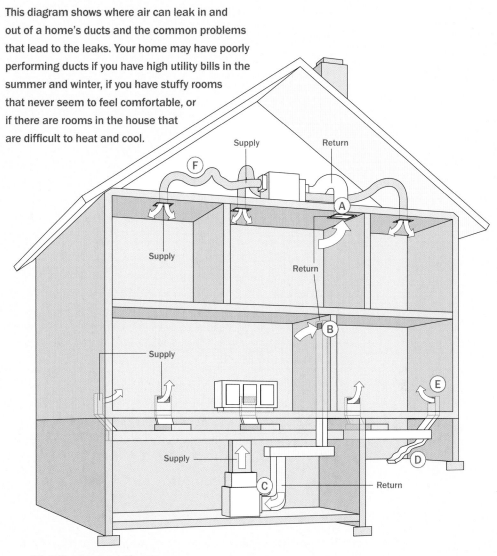

Supply

Return

F

Supply

A

Return

B

Supply

Supply

E

Supply

C

Return

(Adapted from Department of Energy)

Exposed ducts in hot attics, cold basements, damp crawlspaces, and drafty garages are the best places to seal for energy savings, and you should seal and insulate any ducts running through spaces that get hot in the summer or cold in the winter.

A Duct connection leaks
B Return leaks
C Furnace and filter slot leaks; ineffective duct tape seal
D Fallen duct insulation
E Supply leaks
F Restricted airflow because of ductwork kinks

What will this project do for the Earth?

If every home in the United States sealed its leaking ductwork, more than $5 billion in energy costs would be saved per year. This would be the equivalent of removing the carbon emissions of 13 million cars.

Will you need a contractor?

Maybe. Some ductwork is easily accessible, such as ducts in a garage or basement. However, most ductwork is difficult to access, hidden away in floors, walls, and ceilings. For these ducts, you may want to hire a qualified professional who'll be able to reach and seal all ducts.

The most important areas to seal are disconnected ducts, connections between a duct and a vent, and the edges and seams in the furnace or central air unit.

What are the best sources for materials?

Despite its product name, ducts are not sealed with duct tape (or duck tape). In fact, duct tape is perhaps the worst thing you can use on your ducts. The adhesive fails quickly, leaving sticky residue on ducts and not sealing anything. Instead of duct tape, seal ducts using duct mastic. It is a waterproof, flexible sealant that never hardens but dries as a rubbery, stretchy material. This allows the mastic to expand and contract as the duct expands and contracts, all the while maintaining its seal. It works with nearly any type of duct, including metal, fiberglass, and the flexible hose type.

Mastic comes in caulking tubes, tubs, and large buckets. If any of your ducts are visible, the color of the mastic might be important to you. A gray or beige color might be more attractive than basic white. Mastic is typically applied with a brush or caulking gun, though a trowel or even your gloved hand will work. Avoid oil- and petroleum-based mastics and look for the water-based kind. A single gallon is enough to seal 20 to 40 joints.

Applying mastic to ducts

1 Clean the surface and wipe off loose dirt and oil with a dry rag. Make sure duct screws, rivets, or any other fasteners are in place and haven't fallen out. Tighten or replace loose and missing screws. Fit duct pieces together if they have come apart.

2 Seams, gaps, and openings up to $1/4$ in. wide can be sealed with mastic alone. Load up your brush and spread the mastic directly over the joint and at least 1 in. on each side of it. Apply a thick coating

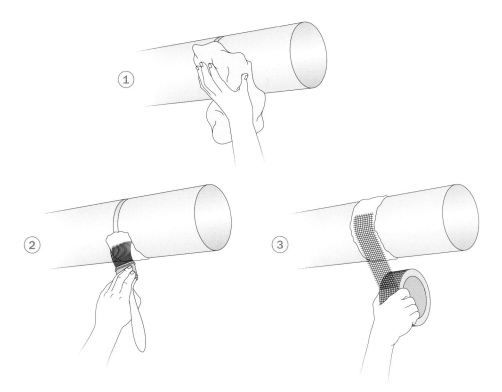

that fills the gap and completely covers the surface of the duct. The finished application of mastic should be about $1/16$ in. thick.

3   For gaps wider than $1/4$ in., use fiberglass mesh tape to provide support over the mastic-coated joint. This mesh is similar to the mesh tape used on drywall joints, but make sure you get mesh specifically designed for use with ducts. Cut the mesh to the length you need and press it into the wet mastic. Apply a finish layer of mastic over the mesh to finish the seal. Allow the mastic to dry for 2 to 4 hours.

Any ducts in hot or cold spaces (attics, crawlspaces, basements, etc.) should be insulated. Wrap fiberglass insulation around the duct after sealing the duct with mastic and allow it to dry.

If your ducts are covered with insulation, remember that insulation alone does not seal air leaks. Look for dirt streaks on the surface of the insulation as a sign of leaks. If the ducts are already covered in insulation, you'll need to remove it to apply the mastic. If your home was built before 1989, the ducts may be insulated with asbestos wrap. Asbestos is a hazardous material best left alone until removed. You'll need to find a licensed asbestos abatement contractor to remove the asbestos.

If you are concerned that you have serious duct leaks, a professional installer can check by using special equipment called a blower door. It measures the exact amount of air leaking from your ducts. It can be used to determine how badly you need the sealing done.

For new construction, consider placing ducts in spaces that are heated and cooled, instead of in an attic or crawlspace. This alone can conserve up to 60% of the energy used in heating or cooling a home. Ducts installed into a conditioned space need to be sealed but not insulated.

A worthy alternative to mastic is mastic tape or aluminum foil tape. Mastic tape consists of a roll of sticky mastic compound peeled from a paper backing and applied to the duct. Aluminum foil tapes are made from a thin aluminum film and applied around the duct like tape. The tapes cannot expand and contract as much as the liquid mastic, so they leak after a few years; since they are not liquid, they don't fill the joints as well either, but they work well enough to be considered.

How much maintenance will be required after installation?

Once sealed, ducts should not require additional maintenance, but ductwork and the mastic should be checked periodically for new cracks or leaks. Dirt streaks are a telltale sign of a leaky duct.

How long will the project take to accomplish?

The process of sealing ducts is not time-consuming. Reaching and accessing them is the hard part. If you choose to do the work yourself, it could be done over several days. A professional could seal all the ducts in your house in a day or two.

KEVIN SAYS I was utterly disappointed when I learned that duct tape was out for sealing the ducts in our house, and mastic was in—requiring a time-consuming, messy, and difficult process of using mastic and getting all sticky and dirty as I try to apply the stuff in hard-to-reach areas. I therefore called a couple of commercial HVAC contractors and, sure enough, metal-backed aluminum sealing tape is specifically designed for the job of sealing metal ductwork.

Now you may have to unwrap some insulation, but the metal tape is designed to withstand cold, heat, moisture, and temperature fluctuation without losing its shape or stickiness. The liquid mastic may perform a little better and last a little longer, but no research I found showed that to be the case, and tape is much easier to work with.

What is the capital cost? $

If you decide to go with the mastic, you can get it and application tools for around $75. Metal-backed aluminum tape is inexpensive and available at most hardware stores. About $35 will buy three 50-yd. rolls.

What financial resources are available? $

A few utilities offer rebates of a couple hundred dollars, but you have to have a contractor come out and certify that your duct leakage has been reduced. The contractor will likely charge you more to do the test than you'll get back with rebate, so it's probably best to go it alone.

What is the monthly cost or savings? $$$$

Benefits vary based on heating and cooling usage and size of the home, but overall, duct sealing is reported to improve heating and cooling efficiency by as much as 30%. If you are a big user of both heating and cooling, you could save upwards of $300 annually. Even the light energy user will likely earn back the cost of the project within the first year.

What is the long-term home value? $$

It's a little difficult to convince a would-be buyer that a $75 improvement is worth paying more for a house, and there is no data to support that a house with sealed ducts appraises at a higher value with a lender. However, if you show prospective buyers utility bills that are lower compared with others in your area, it could go a long way in convincing them that your house is the most energy efficient one on the block.

THE BOTTOM LINE is...

You need to do this. It's a great money saver and a great planet saver. If you like messy, sticky stuff, by all means use the mastic, but at the very least go make it happen with the metal-backed aluminum tape and save big dollars. ■

Modern Duct Sealing

New techniques are now available for instantly sealing every single duct in your home. Products such as Jett-Seal (www.jettseal.com) are sprayed inside ducts and blown throughout the home. They automatically fill in leaks and can be applied in new or existing homes by trained installers. While these products are not yet common, expect to see more of them available in the near future.

RESOURCES

Built Green
www.builtgreen.org

GreenPoint Rated
www.builditgreen.org

Jett-Seal
www.jettseal.com

LEED for Homes
www.usgbc.org/LEED/homes/

National Air Duct Cleaners Association (NADCA)
www.nadca.com

Use healthy paints

10

Green $pecs

Overall Rating

Difficulty

Green Benefits

ERIC SAYS Painting is a simple way to revitalize a room while protecting a home's walls inside and out. More than 850 million gal. of paint were sold last year. But the majority of paints contain harmful chemicals called volatile organic compounds (VOCs) that evaporate into the air (called off-gassing) and create that "new paint smell." VOCs are a major contributor to low ozone (known as smog), but more important, they are a known carcinogen and respiratory irritant.

Nearly every major paint manufacturer offers a line of paints with low VOCs, but being less bad is still bad. Think of it as a low-fat food—there's still fat in it. So look specifically for zero-VOC paint as the guaranteed healthier alternative. By doing so, you can eliminate the main source of indoor air pollution without sacrificing appearance or taking on any additional cost.

What will this project do for your home?

Not only does paint add color and brightness, but painting your walls also protects the wall surface from dirt and fingerprints. By choosing a healthy, zero-VOC paint, the levels of airborne chemicals in a home are drastically lowered.

What will this project do for the Earth?

A gallon of discarded paint in the landfill can seep into the groundwater and pollute 250,000 gal. of drinking water. Paints are the second-largest source of VOC emission pollution (after the automobile). By switching to healthier alternatives, we can eliminate most of this wasteful pollution. Healthy paints also reduce the level of smog in the air, which costs U.S. farmers an estimated $500 million in reduced crop production every year.

Will you need a contractor?

Painting is one of those tasks people either love or hate. But whether you hire a professional painter or do it yourself, the air-quality results will be the same as long as you start with healthful paints.

What are the best sources for materials?

The Home Depot and Lowe's sell their own proprietary low-VOC paints—Fresh Aire Choice® and Olympic®, respectively. Most hardware stores carry paint labeled as low-VOC or zero-VOC, but don't be fooled by flashy names that only sound environmental. Many manufacturers claim to have a low-odor product, but it may not be low-VOC. Read labels to check the exact VOC content. By definition, low-VOC paint contains no more than 50 g/L (grams per liter) of VOCs for flat finishes or 150 g/L for nonflat finishes (eggshell, semigloss, gloss). Also look for Green Seal® or Greenguard labels that certify paints with low chemical emissions. And the Scientific Certification Systems (SCS) is an independent testing agency that verifies the environmental features of any product. Look for the SCS logo as verification of claims about a paint's VOC content.

Indoor air is likely to contain more than 10 times the number of toxic pollutants as outdoor air. During a painting project, that number can increase to 1,000 times the outdoor levels. Paint in a well-ventilated space, wear a mask, and switch to healthy paints containing low or zero VOCs to remove major health risks.

The Greenguard (www.green guard.org) seal indicates a product with low chemical emissions.

Scientific Certification Systems (www.scscertified.com) is an independent testing agency that certifies the environmental claims of manufacturers. Seeing the SCS logo means a manufacturer's claims about a paint's VOC content and off-gassing have been verified.

As a general rule, light colors and flat finishes contain fewer VOCs than dark or glossy paint. Zero-VOC paints cover and apply just as well as standard paint, they come in nearly any color, and healthy paint manufacturers can match the color from any other company.

Paint manufacturers are not required to disclose every chemical in their products. The only way to get a sense of the final ingredients is to ask for a material safety and data sheet (MSDS). The MSDS will list all of the known chemicals and any health hazards posed by the product, and all companies are required by law to provide the information when asked. Take special care to avoid buying any paint containing mercury, cadmium, chromium, or other heavy metals.

Natural paints Aside from VOCs, paints can also contain harmful toxic substances made from energy-intensive and highly polluting sources. New types of finishes, referred to as natural paints, are paint alternatives that use natural binder ingredients, such as linseed oil, soy, pine resin, and citrus, instead of chemicals. Plant-derived compounds are often used for the color pigments. Because the products do not contain any petroleum, the amount of odor is greatly reduced.

Milk-based paints, as the name implies, contain milk proteins mixed with lime, clay, and earthen pigments. Milk paint is sold as a dry powder, and you simply add water to create the paint.

Disposal Never pour leftover paint down a sink or storm drain. Not only is it illegal, but storm drains flow directly into rivers and streams without being treated. Throwing paint into the trash isn't any better. Once in a landfill, chemicals in the paint seep into the ground and water. The greener thing to do is to reserve a little paint for touchups, donate remaining paint, and recycle empty steel paint cans.

How much maintenance will be required after installation?

Zero- and low-VOC paints require no more additional maintenance than standard paints.

How long will the project take to accomplish?

Since they do not contain chemical drying agents, healthy paints tend to take a few hours longer to dry. For any paint job, open the windows and flush the home with fresh air for several days.

KEVIN SAYS In the past, removing toxic VOCs meant a healthy paint's color options were limited. Until recently, commercial zero-VOC paints needed to add VOC-laden pigments that negated their zero-VOC benefits entirely. Lately, however, major brands have added lines of broad-spectrum, eco-friendly paints that allow color freaks like me to go rainbow wild and still save the planet.

What is the capital cost? $

It's difficult to compare the quality and durability of zero-VOC paints because they are relatively new to the market. Zero-VOC brands run higher in price than their VOC-laden counterparts. Premium VOC paint costs around $25 per gal. A midsize house may require as much as 12 gal. for a cost of $300 plus labor. Zero-VOC paints run from $40 to $60 per gal., giving that same house a prelabor cost of $480 to $720.

What financial resources are available? $

There's not much in the way of rebates or credits for healthy paints.

What is the monthly cost or savings?

Since durability isn't any more of an issue for zero-VOC paints than VOC paints, there shouldn't be any impact on monthly expenses.

What is the long-term home value? $

Wall color is purely a matter of taste and preference. Many real estate agents advise clients to repaint colorful interiors white to lighten rooms and make the feel of the home more generic. The theory is that buyers will see past the existing décor and envision their own colors. But, in terms of going with zero-VOC paints, there is little or no data to show that it increases the value of the home.

THE BOTTOM LINE IS...

If it's time for new color, bring it on. There are plenty of healthy paint choices coming down the pike from major manufacturers, which should bring down the price in short order. Even now the extra cost is well worth not having to deal with that horrible paint smell. ▉

Lead Paint Awareness

Although the Consumer Products Safety Commission banned lead paint in 1978, it is a concern when repainting older homes because there could be lead paint under coats of newer paint. Lead paint is a potential source of lead poisoning in children and adults, causing brain damage, nerve damage, and reproductive problems. More than 310,000 children contract related illnesses from lead paint exposure every year.

Dust and loose chips from lead paint are especially dangerous to children and pregnant women. If you're sanding off old paint and are concerned about lead paint exposure, get your home tested. Contact the National Lead Information Center (800-424-LEAD).

RESOURCES

Low- or zero-VOC paint manufacturers:

Anna Sova
www.annasova.com

Safecoat®
www.afmsafecoat.com

YOLO Colorhouse®
www.yolocolorhouse.com

Natural paints:

Auro
www.aurousa.com

Old Fashioned Milk Paint Company
www.milkpaint.com

Weatherize windows and doors

ERIC SAYS The walls of your home are insulated to hold in the conditioned air you spend money to heat and cool, but the largest source of energy loss in your home is your windows. Thin sheets of glass are terrible insulators. While newer windows do a better job of insulating, all windows leak a certain amount of air around their edges.

If you add up the area of all of the cracks and leaks around the windows of your home, it would total about the size of an entire window. Imagine the energy wasted by leaving a window open all winter long! Installing new windows can solve much of this problem, but that can be a big job. Simply weatherizing—sealing the cracks and leaks around your windows and exterior doors—can have an immediate impact on your energy savings and can be completed in an afternoon.

What will this project do for your home?

Unsealed or improperly sealed windows and doors account for more than 50% of the air leaking out of the home. Sealing around windows and exterior doors not only saves energy but also stops drafts and makes it more comfortable for you and your family.

What will this project do for the Earth?

If the windows and exterior doors of every home in the United States were sealed, it would eliminate 6.5 million tons of CO_2 from the atmosphere, the equivalent of taking more than 1 million cars off the road. Enough energy would be saved to provide heating and cooling to nearly 750,000 homes every year.

Will you need a contractor?

Caulking and sealing around windows and exterior doors is one of the easiest, but most effective, things to do to save energy in your home. Anyone can do it themselves, no contractor needed. Get your family to help and make it a chance to bond (pun intended) with the kids.

What are the best sources for materials?

Any local hardware store will carry the basic tools. Look for caulk with environmentally friendly low- or zero-VOC content. Also choose a water-based acrylic product. Press-in-place weatherstripping is a less effective, but easier to install, alternative to caulking. While walking around from window to window, you should also install weatherstripping along the bottom of every exterior door. Threshold weatherstripping is available in a variety of types, typically resembling a thin brush or a spongy gasket. Cut it to match the width of your door, and install it along the bottom edge.

Caulking windows and doors Start by sealing any windows or doors where you can see light coming through the frame. You can find other leaks by holding a wet hand up around the edges of each window and feeling for drafts. Lighting incense sticks and following the smoke is another method to locate drafts. Additionally, a home inspector can conduct a blower door test to identify even smaller leaks throughout the house.

DO THE MATH

If American households saved 10% of energy used to heat and cool their homes, it would amount to 8.2 billion kW saved (see p. 64).

An amount of 8.2 billion kW is equal to 5,888,943 metric tons of CO_2, which is equivalent to the annual greenhouse gas emissions from 1,078,561 passenger vehicles.

1 car emits 5.46 metric tons of CO_2 per year

1 metric ton = 1.10231131 short tons

5,888,943 metric tons = 6,491,448 short tons

See http://www.epa.gov/cleanrgy/energy-resources/calculator.html for more info.

Air Leak Locations in the Home

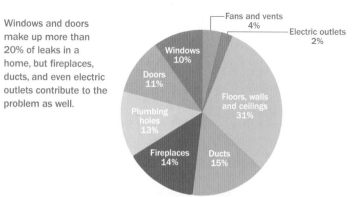

Windows and doors make up more than 20% of leaks in a home, but fireplaces, ducts, and even electric outlets contribute to the problem as well.

Fans and vents 4%

Electric outlets 2%

Windows 10%

Doors 11%

Floors, walls and ceilings 31%

Plumbing holes 13%

Fireplaces 14%

Ducts 15%

(Adapted from data courtesy of U.S. Department of Energy, Energy Savers)

Cracks occur around all windows, letting conditioned air leak out and outside air enter your home. Caulking around windows and doors prevents these drafts and greatly improves the energy efficiency of your home.

Outside your home, go to each window and door and use a metal scraper to clean the joint between the window and the wall. Remove any flaking paint or dirt. Inspect areas where different building materials meet—between the chimney and the siding, around the dryer vent, and along the bottom edge of the siding. These areas benefit from caulking.

Caulk any leaks with a 1/8-in. bead between the window frame and the exterior siding. The goal is to fill in any gaps where air and water could leak into your home. Use an ice cream stick (or even a gloved finger) to smooth the bead of caulk. Wipe off any excess to keep it from oozing or dripping. Periodically wipe the tip of the caulk tube to ensure a neat bead, and allow the caulk to dry for at least a day before painting.

How much maintenance will be required after installation?

Caulking cracks over time, so check joints annually. If you plan on painting in the near future, it's best to caulk beforehand.

How long will the project take to accomplish?

You could easily fill an entire afternoon caulking around the doors and windows of your home. It's best to apply the caulk on warm, dry days, so wait for a nice day to tackle this project.

KEVIN SAYS Whether it's caulking windows or weatherstripping the doors, weatherization has become a national priority. But some people have concerns about sealing every hole in the house, worried that enough fresh air won't get in.

According to the American Society of Heating, Refrigerating and Air-Conditioning Engineers (ASHRAE), one-third of the air in a home needs to be replaced every hour to keep it fresh and healthy. Even after carefully caulking and weatherstripping, the air in a typical home will be replaced up to eight times every hour, so the only things getting stale are the excuses for not weatherizing.

What is the capital cost? $

Eco-friendly caulk with no VOCs is a little tough to find, but a couple of Internet searches should get you there, and the price at $6 to $8 per tube isn't much different than most colored sealers. Figure that six to eight tubes at a total cost of no more than $65 should be enough to seal a house with 15 to 20 windows. You might spend another $40 to $75 to weatherstrip your doors. All in all, the job is done for less than $150.

What financial resources are available? $$$

Most weatherization programs offered by local governments focus on window replacement or adding insulation. However, a few municipalities such as Azusa, California, have put forward a small amount of money for caulking, like offering up to $25 in rebate, but not all cities do this. Check your local municipality.

The federal government has apportioned funds for weatherizing homes through the Low Income Home Energy Assistance Program (LIHEAP). The government will provide as much as $2,800 to help weatherize homes (including insulation) for families who rent or own. The states administer the program and calculate eligibility based upon their own criteria; call 866-674-6327 for more info.

Another way of weatherizing the home is with plastic shrink wrap. It's easy to install and holds in the temperature, but is clearly not practical if you want to open your windows.

What is the monthly cost or savings? $$$

Weatherizing doors and windows will definitely save you money, but just how much will vary. Factors include the age of the home, the types of windows and doors, and personal climate and temperature habits. Owners of older homes with no previous weatherization could save as much as 15% to 20% of their energy bill. Newer homes will likely see less savings since much of the sealing has already been done. Either way, your savings will pay for the cost of the project in a few months.

What is the long-term home value? $

Although there is no specific data to show that caulking and weatherstripping actually increase a home's value, nicely caulked windows and weatherstripped doors will

In addition to weatherizing and caulking, tinted films can be applied to windows to block ultraviolet (UV) light from a home. UV light can bring in unwanted heat during the summer months.

DO THE MATH

The energy lost through leaky windows and doors in U.S. homes is enough to heat and cool almost 750,000 homes every year. Here's how it is broken down:

There are 111,000,000 households in the U.S., which use 82 billion kW each year for heating and cooling.

A 10% energy savings (from sealing leaks, let's say) would amount to 8.2 billion kW.

In 2006, the average annual residential electricity consumption was 11,040 kWh.

Divide the potential energy savings by our average monthly energy consumption (8.2 billion kW / 11,040 kWh) and you'd have enough energy to heat and cool 742,753 homes. At 12.4¢ per kW, 8.2 billion kW of energy cost $1,016,800,000, which could be saved.

See http://www.eia.doe.gov/emeu/recs/recs2005/c&e/spaceheating/pdf/tablesh1.pdf for more information.

usually impress a buyer, particularly in an area with extreme weather. As long as it's all been done in an aesthetically pleasing manner, buyers should appreciate the fact that it's been taken care of and you can demonstrate low energy bills.

THE BOTTOM LINE IS...

Do your part for the planet and economy by sealing around the house. Leaks can cost U.S. homeowners more than 1 billion dollars a year in wasted energy costs, and who wants the utilities to get all that extra money? The hardest part to this project is getting the caulk bead to look nice, so hire a handyman or figure out who has the steady hand in the family and get him or her to do it. If you decide to do the work yourself, take your time and do a thorough job so you don't end up half-caulked. ■

Install solar pool heating

ERIC SAYS On a hot day, there are few things as refreshing as a dip in the pool. Having a pool can be great, but pools require ongoing care and maintenance, and they also consume an incredible amount of energy, most of which goes into running the pool water heater.

At a time when power is growing scarce and more expensive, heating pool water might seem excessive, but there's a better way to do it. The sun provides abundant, unlimited, and free energy. More sunlight energy hits the Earth each hour than the world's population consumes in an entire year.

Solar heating is simple, reliable, and has been used for decades. Since solar heaters only need to warm the water slightly, solar pool heating systems are less complicated than full domestic solar water heating systems. This system can cut your pool heating by anywhere from 50% to 100%.

What will this project do for your home?

Pool heaters keep the water at a constant 78°F to 82°F, a simple task for a solar heater (compared with the 120°F to 140°F of a domestic hot water heater). With no moving parts and practically no maintenance, solar pool heaters can also remove the hassle of heating a pool as well as potentially eliminate all the energy costs.

What will this project do for the Earth?

Standard pool heaters use electricity or natural gas to slowly heat water. Heating the approximately 8 million pools in the United States consumes enough energy to power more than 688,000 homes per year. Nearly 7.6 billion kilowatts are used each year for pool heating, contributing to carbon emissions and our dependency on fossil fuels. By saving this energy, you'll help slow the effects of global warming.

Green $pecs

Overall Rating

Difficulty

Green Benefits

In the U.S., 7.6 billion kW are used every year for pool heating (about 2,300 kWh for each of the 3.3 million pool heaters out there). In 2006, the average monthly residential electricity consumption was 920 kWh or 11,040 kWh per year.

7,600,000,000 kW / 11,040 kWh = 688,405 homes that could be powered with the energy used to heat the country's pools

See http://www.eia.doe.gov/emeu/reps/enduse/er01_us_tab1.html for further information.

Will you need a contractor to complete this project?

Installing a solar pool heating system can be a bit complicated. Making secure plumbing connections from your roof to the pool filter can be tricky, and bolting large panels to a roof can be dangerous. Leave the work to the experts.

Solar pool heating systems are typically installed by pool installation companies. Ask for (and check) references and make certain they are both licensed and insured (bonded) in case they damage your roof or someone is hurt during the installation. The contractor will prepare an estimate by visiting your home and looking at the roof and pool. You will need a large south- or west-facing area on your roof to hold the large panels of the heating system.

What are the best sources for materials?

When choosing a solar pool heating system, make sure it's certified by the Solar Rating and Certification Corporation (www.solar-rating.org). Solar pool heating systems can be installed on nearly any type of roofing. Unlike solar electricity panels, the roof attachments for solar pool heating are minimal. The system will need to be drained during freezing months, but unless you're a polar bear, you're probably not using the pool then anyway.

Although a solar heating system is usually placed on the roof of a house, it can also be installed on a garage or shed—or even on the side of a hill or in the yard—if it receives enough sunlight and is close enough to the pool equipment. In general, the size of your collector will be around half the size of the pool surface area. This varies based on a number of factors, including the climate and average local temperatures, the amount of sun striking the roof, the angle and orientation of the roof, whether a pool cover is used, and the number of months a year the heater will be used.

Since solar heaters require sunlight to work, they will not perform (as well) on cloudy or cold days. A standard gas or electric heater can be used as a backup. If you live in a place with clear skies and sunshine during the heating months, you may be able to avoid installing a conventional heater altogether. Even if you do occasionally need to turn on your gas or electric heater for cloudy days, your solar heater will provide a large amount of free energy.

By far, the best accessory for your new solar pool heating system is a pool cover. Evaporation is the largest source of heat loss for a pool.

Solar Pool Heater

A solar pool heating system is surprisingly simple. Dark colors absorb heat (similar to how a blacktop road gets hot in the summertime). Large black fabric panels—called collectors—are stretched over a sun-drenched roof. A supply pipe feeds cool water from the pool filter into the bottom of the collector. Sunlight heats the collectors, warming the water. As the water heats up, it rises to the top of the collector. Another pipe along the top of the collector carries the warm water back to the pool.

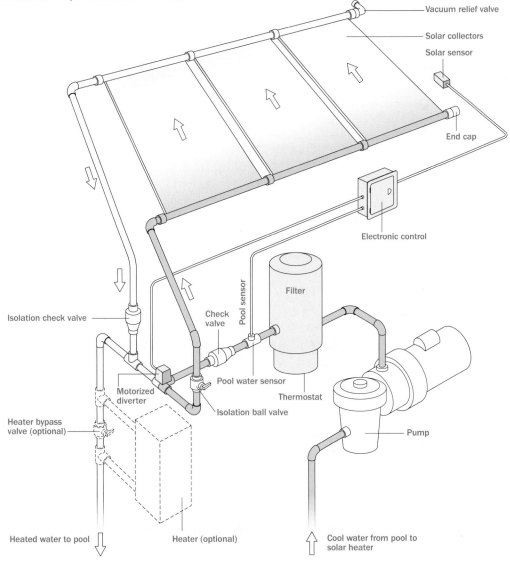

Vacuum relief valve

Solar collectors

Solar sensor

End cap

Electronic control

Filter

Pool sensor

Check valve

Isolation check valve

Thermostat

Pool water sensor

Motorized diverter

Isolation ball valve

Heater bypass valve (optional)

Pump

Heated water to pool

Heater (optional)

Cool water from pool to solar heater

Using a pool cover can eliminate 30% to 50% of the evaporation as well as reduce the need for chemical use by 30% to 60%, prevent dirt and debris from falling in the pool, and keep your pool warm and lower the strain on the heater.

How much maintenance will be required after installation?

Solar pool heating is automatic, requiring almost no maintenance or operation. A thermostat automatically sends water from the filter to the collector. Once the preset temperature is reached, or when the sun goes down, the process stops. A conventional gas or electric heater can make up the difference when needed. If you decide to move to a new home, you can take your solar pool heating system with you. (Just be sure to warn your real estate agent!)

How long will the project take to accomplish?

Installation can be completed in a day, but a solar pool heating system should last for at least 20 years.

KEVIN SAYS Growing up with a pool in the '70s during the energy crisis, I remember my dad always looking for ways to heat it cheaply. We had a big solar cover because utility costs were enormous. I know Dad and I would have both been happier with an automatic solar system to heat the pool. However, I'm not sure the savings would have warranted the cost, since before the cover we simply used the pool only in the warmer months.

What is the capital cost? $$$

The size of a pool has a lot to do with the cost of the solar heating system you need. There are a variety of suppliers out there, but the bulk of them charge from $500 to $2,000 for an uninstalled system. You can add fancy timers and accessories that can drive the cost up another couple thousand. You could be looking at an additional $1,000 to $3,000 to have someone install the system, bringing the total cost to anywhere from $1,500 to $7,000. Attempting the installation yourself lowers costs but may not be worth the risk if you aren't comfortable working on the roof. A solar cover is substantially cheaper at a cost of roughly $200 to $500 for most pools.

What financial resources are available? $$

The federal government specifically excluded pools and spas from the credits they provide for solar heating. Several states such as Oregon, Florida, and Arizona, however, understand the need for pool lounging and have provided various tax credits with certain restrictions. Check your state's website to find more information or ask the manufacturer of the system you are interested in buying.

What is the monthly cost or savings? $$

Solar heaters can save upwards of 70% of the cost of a gas or electric pool heater. Savings will vary based upon usage and the size of the pool but could range from $50 to $250 per month during the time of the year when the pool is usually heated. There will be no savings during the months that you aren't using the pool.

What is the long-term home value? $

In some areas, an in-ground pool is a luxury. In places like Phoenix and Los Angeles, pools are virtually mandatory for decent resale. The ability to market your house with a year-round heated pool at a low cost will certainly grab attention from buyers. It won't bring a lot of money to the selling price but should be enough to recoup the investment.

THE BOTTOM LINE is...

Solar makes sense regardless of how you use the pool or where you live. If you are an infrequent user, go the cheap route with a solar cover. It will keep the leaves out and save money on hiring a pool cleaner. If you're a daily swimmer, you'll love those extra months of swimming that a solar pool heater gives you, and you can revel in being environmentally friendly, too. ▦

RESOURCES

Solar Rating Certification Corporation
www.solar-rating.org

Plant native species and skip the sprinklers

Green $pecs

Overall Rating

Difficulty

Green Benefits

ERIC SAYS Some 54 million Americans get out every weekend to mow and water approximately 20 million acres of lawn. Most lawns are watered with old-fashioned sprinklers, spraying water on the top of the grass. Nearly 80% of that water evaporates and never reaches the grass roots. Still, we use anywhere from one-third to two-thirds of our domestic water on our lawns—while billions around the world go thirsty.

One solution is to stop having lawns of grass altogether and to grow native plant species instead. You can reduce the need for all the mowing and watering and still have a landscaped home—but one that thrives in your region's climate. But if a grass lawn still holds an appeal, don't water with a sprinkler. Instead, install a drip irrigation system that uses much less water—bringing it directly to the roots where it belongs.

What will this project do for your home?

During the summer, 30% to 60% of the average home's water ends up on the lawn. A drip irrigation system can cut that use by 80% and requires much less maintenance for your lawn. Replacing the lawn with a native rye grass (or other appropriate plants) could eliminate all watering except on the hottest of days.

What will this project do for the Earth?

Global water consumption has nearly doubled since World War II, leaving more than a billion people without access to clean drinking water. Due to global warming, ocean levels continue to rise. The Pacific Ocean below the Golden Gate Bridge has risen 8 in. in the past century. Our fresh water supplies are at risk, and yet more than half of U.S. water use in urban areas is for landscape irrigation.

Will you need a contractor?

Some people might enjoy the idea of digging in their lawn on their knees for hours at a time. If you have a green thumb, this project will be fun. Otherwise, an experienced landscape contractor can help source the plants and irrigation system, and save your knees.

What are the best sources for materials?

Your local plant nursery or gardening center can help advise you on the plants that are appropriate for your climate. Look for plants that can live off the local rainfall and do not need additional watering or pesticides to flourish. Plants can be added in phases and over time as your budget allows. This could also be a good opportunity to make part of your yard into a garden to grow vegetables and herbs.

If watering and irrigation are still needed, a drip irrigation system is the answer. A drip system can be installed above or below ground.

Drip Irrigation System

The parts of a drip irrigation system are pretty simple: a sensor, a pump, and hundreds of feet of flexible tubing. The system saves up to 90% of the water that a traditional sprinkler system uses.

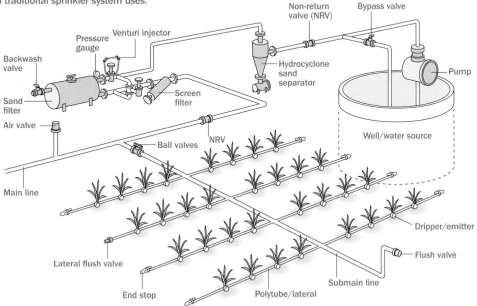

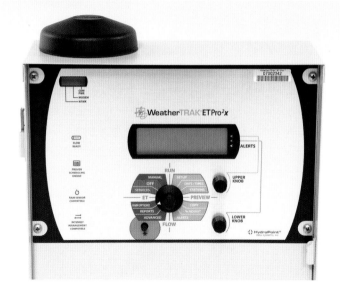

A weather-tracking system is a convenient and worthwhile accessory for your watering system. This small box checks the local weather and automatically shuts off the watering system in the event of rain. The investment could pay for itself in a year or two from water savings. Alternatively, a simple rain switch that engages when it rains is a more affordable solution, depending on your system.

Although more complicated to install, a below-ground drip system conserves more water since the water doesn't get a chance to evaporate—it is installed a few inches below the soil near the roots. Flexible polyethylene tubing slowly drips water to the roots as needed, eliminating overspray and overwatering. The consistent watering encourages healthier plant growth and makes the plants more resistant to hot weather.

If you're planning on digging up or regrading your yard, that would be the ideal time to install a drip irrigation system. Otherwise, install a surface drip system. You'll save money and headaches by installing the entire system at once. Installing the system in phases increases the chances of leaks or damaging the plants and irrigation already in place.

Since you're digging up the yard, you'll need to watch out for utility easements, buried water, and gas and power lines. If you're unsure where they might be, contact your local utility for the locations of these items.

How much maintenance will be required after installation?

Drought-tolerant plantings require only minor care and maintenance like watering during an unusually dry spell. Native plants are already attuned to their climate and require equally little upkeep.

The drip irrigation system should be checked every couple of years for breaks in the line. Of course, any breaks would start to show up as dead patches of lawn and alert you to the fact that something is wrong. The biggest worry comes from digging around in your yard and potentially cutting the tubing.

How long will the project take to accomplish?

Planting native and drought-tolerant plants can be done over a few weekends, while replacing an entire lawn with hardy rye (or other appropriate plantings) could take a couple of weeks. A drip irrigation system could be installed at that same time; it takes about a day to lay the lines.

KEVIN SAYS When my wife and I bought our home in Connecticut, we were both enamored and intimidated by the elaborate woodland garden and natural lawn that had been carefully tended for 30 years by the previous owners. Coming from California, where everything was on automatic sprinklers and required a regular visit by the "Mow, Blow, and Go Guys," we were sure we were in for a lot of work and most likely would ruin the landscape with our black thumbs. Several years later, the gardens still look fantastic. This is the advantage of planting native plants and grasses. It's really quite simple. Foliage that grows naturally in an area requires little maintenance to thrive. That means you spend less time, less money, and less effort maintaining your property.

What is the capital cost?　$$

Drip systems aren't particularly expensive. Depending upon the density of your planting and the kit chosen, you can expect to spend around $0.30 to $0.80 per sq. ft. of space if installing it yourself. That means an irrigation system for a 20-ft. × 20-ft. area would cost around $120 to $320. A landscaper or gardener might charge you another couple of hundred bucks to install it. If you are re-landscaping the whole yard from scratch, you might expect to pay $1,000 to $3,000 to have all the systems designed and installed. Still, it would be less than half the cost of the trench digging and hardware purchases associated with an extensive sprinkler system.

Regarding local plants, you may have some difficulty finding a large variety at the big-box stores, but local nurseries can usually get you a broad selection. The cost may run a touch more per plant, but you get expert advice and high-quality plants.

Choosing less lawn and more native plants can deliver immediate savings. The savings per acre of plants versus sod can be as much as

$3,000 to $4,000 when landscaping from scratch. Natural native grasses can be more easily planted by seed, which can save even more. The biggest expense would be the total replacement of a nonnative yard, but if you are on a tight budget you can always amortize the cost by budgeting ahead for 6 to 10 replacement plants per year. Pick concentrated areas and then cap off the sprinklers in that area to start saving.

What financial resources are available? $

While there is not much money from federal or state resources, local water districts in drought-affected states such as California have limited funds available for drip irrigation system reimbursements. Some allow a few hundred dollars in credit or will even provide the systems themselves. Checking with your local water district can be fruitful. Many local nurseries are getting on the Green bandwagon and use native plants as loss leaders for their marketing, so careful shopping may yield savings.

What is the monthly cost or savings? $$$$

In 2003, the City of Santa Monica, California, did a study comparing the maintenance of a traditional lawn with one that had native plants and

Tips to Reduce the Impact of Your Lawn

Installing a drip irrigation system and planting new plants is not an option for everyone. If you are keeping your existing lawn, here are a few tips to help reduce water and energy usage:

1 **Water at night.** The heat of the sun speeds evaporation and prevents most of the water from reaching the roots. By watering at night, you're giving the water a better chance of getting where it needs to go.

2 **Get some exercise.** Why pay a monthly fee to join a gym when you can get exercise in your front yard? Switching to a push-type mower (instead of a gas one) helps cut 80 lb. of greenhouse gases per lawn per year. Lawn mower engines are inefficient and polluting, producing up to 5% of the nation's air pollution.

3 **Don't cut it so short.** Oddly enough, most grass is cut shorter than it should be. If your grass is too short, the blades cannot shade the soil, which increases the need for extra watering. Keep it slightly longer to help your lawn survive.

4 **Grass-cycle.** Leave loose clippings on your lawn after mowing. They will decompose and help fertilize the soil naturally. Plus, this grass-cycling means less work for you.

5 **Skip the fertilizer.** Spraying poisonous chemicals onto your lawn to make it greener is no way to go. Why risk the health of your family and pets when natural methods work just as well? If you want a greener lawn, try the tips above instead.

efficient irrigation. The native garden used seven times less water and required 50% less cash to maintain than the traditional lawn. In addition, studies have shown that natural seed planting versus manicured lawn provides almost $150 per month per acre in savings when considering mowing, fertilizing, replanting, and aerating.

Natural gardens can save upwards of 50% of a water bill if you are currently caring for a large lawn. That could be more than $100 per month. For those of you doing the weekend tractor ride, switching over to native plantings gains you another $10 a month in lawn mower fuel cost savings, plus you won't have to replace the mower every few years. And our time is worth plenty, and while I like gardening, I don't want to spend all my time doing it. What's a weekend really worth?

What is the long-term home value? $$$

There have been a couple of studies examining the aftermarket value of native gardens in resale homes versus traditional lawn-based landscaping. The Sacramento Tree Foundation showed that each large front yard tree adds 1% to 10% to sales prices, and a 2005 Michigan study found that people were willing to pay more for well-designed yards—including native plants—than they would for yards with lawns.

Bank appraisers have long given value to mature landscaping and lush gardens over sparse landscaping. Fancy landscaping is almost never given dollar-for-dollar value versus cost, so reducing the investment for a more natural look can work in your favor. Even prickly cactus gardens in desert areas, when done well, have been shown to attract buyers over a basic lawn and flowerbed offering. Keep in mind that unsightly and unappealing landscaping, regardless of cost and maintenance, will decrease the value of your home, so hire a landscape designer if you aren't sure where to start.

THE BOTTOM LINE is...

Gardens and lawns are a matter of taste. If you absolutely need the manicured lawn look for your home, you and the planet will both pay a premium in dollars and energy. But think about what first appealed to you about the place you chose to live. If you embrace the local scenery and foliage, you can save a ton of money on your yard upkeep—and even gain more time to enjoy it. ■

RESOURCES

WeatherTRAK
www.weathertrak.com

Buy a water filter instead of bottled water

Green $pecs

Overall Rating

Difficulty

Green Benefits

ERIC SAYS We are bottled water addicts. In the United States, we consume 6.9 billion gal. of the stuff a year, filling some 30 billion of those little plastic bottles. It takes 1.5 million barrels of oil to produce all of that plastic, enough to fuel 100,000 cars for a year. Ironically, it takes three to four times the amount of water to make the bottle than actually fits into it. Sadly, only around 20% of all those billions of plastic water bottles are recycled.

We spent more than $15 billion in 2008 on bottled water, assuming it was cleaner or healthier. But at least a quarter of bottled water is just filtered tap water. Check out the National Tap Water Database (www.ewg.org/tapwater/yourwater/) if you are skeptical. Recent reports from Case Western Reserve University also found that 15 of 39 samples of bottled water contained twice the bacterial count of Cleveland tap water. And bottled water lacks fluoride, increasing the chance of tooth decay. In the U.S., tap water is regulated by the stringent Environmental Protection Agency (EPA), while bottled water is regulated by the lower standards of the Food and Drug Administration (FDA).

You already have an endless, cheap, and clean source of water being delivered directly to your home. It's coming from your kitchen faucet. Add a water filter, and you'll be rid of any lingering concern over the quality of your tap water. Companies like Brita® and Pur® produce filtering pitchers and single-faucet filters. But we can go one step further—a whole-house filter purifies every drop of water coming into your home.

What will this project do for your home?

Filtering reduces the amount of chlorine and particles in water, improving taste and odor. A whole-house filter eliminates the hassle and expense of buying bottled water, and filters the water you cook with, drink, bathe in, and wash your clothes with.

What will this project do for the Earth?

Every ton of plastic bottles produces three tons of carbon dioxide, the major source of global warming. Bottled water produced 2.5 million tons of CO_2 in 2008, equal to the emissions of 415,000 cars. Distributing water around the country puts 37,800 18-wheelers on the road every week. Peter Gleick, director of the Pacific Institute, says the true cost of bottled water is "like filling up a quarter of every bottle with oil."

Of the 80% of bottles that don't get recycled, many of them end up in our oceans. The Great Pacific Garbage Patch—a floating island of plastic trash twice the size of Texas, located in the Pacific Ocean between California and Japan—is composed largely of old plastic bottles. Installing a whole-house water filter (and giving up bottled water for good) reduces this impact and makes a substantial difference in our environment.

Will you need a contractor?

If you're comfortable with valves, pipes, and plumbing, this is a project you can do yourself. If the thought of shutting off the water to your home is an intimidating prospect, contact a good plumber to help.

What are the best sources for materials?

A wide variety of whole-house filters are available from plumbing-supply stores; many can be purchased online. The filter is typically installed next to the water meter, where the water line comes into the home. Since water passes through the filter, no water is lost or wasted during installation or otherwise. Some models offer a small storage tank to provide an uninterrupted supply of fresh water. A whole-house water filter is not recommended for outside installation.

For times when you leave the house, purchase a reusable water bottle, such as the ones made by Sigg™, Nalgene®, or Klean Kanteen™. These bottles are washable, durable, and safe. They make the perfect companion to a whole-house water filter.

How much maintenance will be required after installation?

Unlike their tiny pitcher counterparts, whole-house water filters are maintenance free. Most models are typically warranted for 10 years and more than half a million gallons. If you're worried about the expense and hassle of changing filters, several companies offer filter-less or self-cleaning filters that work just as well.

How long will the project take to accomplish?

A plumber can install a whole-house filter in a couple of hours. Remember that your water main will need to be shut off during the installation, so pick a time when no one in the family needs to shower or do laundry.

Once you've installed a water filter at home, bring your own "bottled" water with you wherever you go in a reusable and durable container.

KEVIN SAYS There's no question that tap water, even with a filter, is much cheaper than bottled water. Even gasoline at $4 per gal. is still less expensive than typical bottled water from the store. So what are we paying for? Taste and convenience are the two big attractions. Filters actually do a great job of solving the taste problem and can even solve the convenience issue at home. The savings can be decent, that is if you aren't already satisfied with the taste from the tap.

What is the capital cost? $

Big-box stores have single-faucet and whole-house filters ranging in price from about $30 to $200. Installing a whole-house filter may require plumbing help if you're not handy, so you might have to spend another $50 for a good handyman. Either way the cost is low.

What financial resources are available? $

No government help here, but a little coupon watching in the Sunday newspaper circulars can yield discount coupons on filters and faucet systems from companies such as Brita.

What is the monthly cost or savings? $$

If you are a bottled water junkie, then you'll be reasonably happy with your savings results. A typical replacement filter for faucets is $20 to service 100 gal., or $0.20 per gal. as compared with about $1 per gal. for the cheapest bottled water at discount stores. If your family consumes 2 gal. of water a day on average, the savings would be $48 monthly.

A whole-house filter is cheaper per gallon at $40 for 6,000 gal. or less than a penny per gallon, but much of that water is going to showers and toilets, so it may not be much cheaper for drinking. Most filters have a six-month life, so in our 2-gal.-per-day habit, the cost is about $0.11, and you get all the other clean water free.

The water and energy cost of using your tap is negligible compared with the savings of not buying bottled water. The only way in-home filtering costs more is if you are already drinking regular tap water.

What is the long-term home value? $

No appraisers or studies I could find would give any value to basic water filtration systems.

THE BOTTOM LINE IS...

If you are already tap-water happy, then there is no reason for you to worry. For the rest of us who can't give up the bottle, it's time for rehabilitation. Water filtration is cheap and easy for any home and will certainly make a positive impact on the planet. I was a bottled water junkie and have since become very happy with chilling filtered pitchers in my home. But outside the house is a little tougher. Carrying a canteen can be a hassle, but it's a lot cheaper than buying bottled water on the go. ■

RESOURCES

Klean Kanteen
www.kleankanteen.com

Nalgene
www.nalgene–outdoor.com

National Tap Water Database
www.ewg.org/tapwater/yourwater

PuriTeam
www.puriteam.com

Sigg
www.mysigg.com

Install a clothesline

Green $pecs

Overall Rating

$ $ $

Difficulty

Green Benefits

ERIC SAYS Modern living is all about convenience. Life is busy enough, so any technology that can save time is a welcome thing. In the good old days before mechanical clothes washers, people scrubbed clothing by hand on a washboard. It was backbreaking, tiresome work. Hanging the clothes to dry was the easy part compared with the knuckle-scraping efforts of washing.

Today, 80% of households have a washer and dryer, but this convenience comes at a price. Electric clothes dryers eat up 10% of a home's energy. Each load of laundry gives off around 5.6 lb. of carbon dioxide. That adds up to more than 2,000 lb. of CO_2 per year just from drying clothes. A solar-powered clothes dryer is a highly energy efficient way to dry your clothes. Also known as a "clothesline," this idea has been around for centuries and provides an affordable, easy alternative to the high cost of clothes-drying convenience.

What will this project do for your home?

The clothes dryer is the third largest energy hog in a home, after the refrigerator and washing machine. Washing machines are likely here to stay, but switching to a clothesline is a small choice with a huge impact.

Clotheslines help clothes last longer, sunlight naturally bleaches and disinfects clothes, and you save energy and shave at least $25 off your monthly utility bill by going the solar power route. And hanging your clothes can be a great family bonding experience. Remember hanging clothes with Grandma?

What will this project do for the Earth?

There are more than 88 million clothes dryers in the United States. If everyone tried using a clothesline for just six months, it would save

more than 3% of our global warming emissions. Longer use would result in even greater savings.

Will you need a contractor?

Tightly tying a long length of rope in your yard is enough to create a simple clothesline. Although one person can easily do it solo, it helps to have another person lend a hand. If possible, orient the line to run from east to west to maximize the clothes' exposure to sunlight.

What are the best sources for materials?

Most hardware stores carry rope specifically designed for clothesline use. They come in a variety of types: retractable lines, pulley systems, and self-tightening models. Although vinyl clotheslines are durable and waterproof, they are made from plastic. If you can, go with a nylon or cotton rope clothesline. It'll last for years and works just as well.

If you live in a community with a Homeowners Association (as some 60 million of us do in the United States), there may be rules about installing a clothesline. Usually, they restrict placing a clothesline anywhere visible from the street. Check your local regulations and remember that some rules are made to be broken. Florida is considered a "right to dry" state, where local law supersedes any rules against the use of clotheslines, and other states are considering similar legislation. Check out Project Laundry List (www.laundrylist.org), a nonprofit organization fighting for the right to dry around the country.

Rules or no rules, choose a spot for your line that's not potentially offensive to neighbors or blocking anyone else's view. If you have a small space, or no easy way to hang a line, a drying rack might be the best choice. Thousands of different interior clotheslines and drying racks are available (see Resources on the facing page).

If you decide to keep that old energy hog of a dryer, be sure to clean the lint filter before every use, which increases the airflow and energy efficiency of the dryer. A clogged lint filter forces your dryer to use up to 30% more energy.

How much maintenance will be required after installation?

Clotheslines may sag over time and require tightening. Otherwise, they require no maintenance.

How long will the project take to accomplish?

Installing a clothesline can be done in a matter of minutes. Hanging the clothes should take only 5 to 10 minutes and provides an opportunity to talk with your spouse or kids while doing the planet a favor.

Drying your clothes inside your home helps cool it in the summer and humidify it in the winter. Make sure you hang laundry in a well-ventilated room to prevent any mold problems. A bathroom, basement, or garage is a great spot for an interior line.

KEVIN SAYS Isn't it great that sometimes the best technology is no technology at all? But there are two issues that make me think twice about this planet saver. The first is time. Busy people need to know they can have clothes ready at a moment's notice for work or going out. Space is the other factor. Unless the weather in your area is perfect and predictable, you are always going to give up a room to the clothes rack. But if you can manage the time and space problems, it's an easy money saver.

What is the capital cost? $

The cost of a clothesline ranges from $5 for a rope and hooks and up to $35 for a simple retractable outdoor clothesline. You can spend as much as $200 for something freestanding for indoor use. All in all, even the most expensive approach is still less than half the cost of a new electric or gas dryer.

What financial resources are available? $

Other than the occasional discount coupon in your weekly neighborhood hardware circular, no tax credits or rebates appear to be available for hanging your clothes on the line.

What is the monthly cost or savings? $$$

Saving money is the best part of air-drying clothes. Depending upon usage, your monthly energy costs for a dryer could range from $50 to $100 per month. The bigger the family, the more laundry done, which means more savings. Additionally, there's the savings on wear and tear to the dryer, so you likely won't have to replace it at a cost of $300 to $750 every 10 years or so.

What is the long-term home value? $

The main property value issue with clotheslines is one of looks. If you are using a portable indoor line or retractable system, there's no impact either way. If you have created a large, permanent, cemented monstrosity that takes over the yard, it most likely will have negative impact. Realtors frown upon clotheslines in prominent places because most people do not want to temper the excitement of new home buyers with ho-hum thoughts of laundry labor.

THE BOTTOM LINE IS...

If you have the space indoors or live in a mostly dry, warm climate, a clothesline is a cheap and decent option to save the planet. You'll have to be patient or plan ahead, but it will be worth the money and you don't have to listen to the nagging sound of the dryer running. So the clothes and towels might be a little stiff—at least they won't shrink. ■

RESOURCES

Cord-O-Clip
www.cordoclip.com

Project Laundry List
www.laundrylist.org

Urban Clotheslines
www.urbanclotheslines.com

Recycle and compost your trash

Green $pecs

Overall Rating

Difficulty

Green Benefits

ERIC SAYS The average American produces 4.6 lb. of trash every day, which totals up to 251.3 million tons per year for all of us combined. That's a lot of trash, but consider this: We produce twice as much trash today as we did just 30 years ago, and we produce twice as much trash as most other countries. Only about one-third of that trash gets recycled or composted, while 55% ends up in the landfill.

The amount of trash we bury in landfills has doubled since 1960. Landfills pollute our water, take up enormous amounts of space, and (surprise, surprise) no one wants to live near them. Most people don't realize the biggest problem with landfills is the emissions they generate, namely methane and carbon dioxide gas, which contribute to global warming. By composting and recycling, we can reduce the trash in landfills and do long-lasting good for our environment.

Recycling is not perfect. We've all heard stories about cans intended for recycling ending up in the landfill. Or how the only ones profiting from recycling are collection companies. False rumors that claim it takes more energy to recycle than is saved persist as well.

But look at what we recycle: aluminum cans, plastic bottles, newspapers. These were not initially designed to be recycled, but technology allows us to do it. It does take energy to recycle them, but the energy saved is far greater, and all it takes from us is separating our trash into piles, and putting food scraps and yard clippings into a compost bin.

What will this project do for your home?

Recycling teaches us the importance of protecting our resources and environment and, with composting, can greatly cut down the amount of trash in your home and reduce the number of times you might have to empty the can in your kitchen.

Composting also benefits your lawn and garden. Spread nutrient-rich compost instead of store-bought mulch, and you'll see a change in your plants' appearance within weeks.

What will this project do for the Earth?

Methane and carbon dioxide emissions from landfills are responsible for 3.8% of the U.S.'s contribution to global warming. Last year, recycling reduced the country's carbon emissions by 54 million tons, equivalent to taking nearly 9 million cars off the road.

Beyond global warming, recycling and composting save energy. Nearly 10% of U.S. oil consumption goes into making plastics. By recycling plastic, we save oil. According to the EPA, for every soda can you recycle, enough energy is saved to power a television for three hours.

Using recycled materials also saves energy. Processing recycled glass uses only two-thirds of the amount of energy that making raw glass requires. Producing recycled paper uses 60% less energy than making new paper. Every ton of recycled paper saves 7,000 gal. of water and 31 trees from being consumed.

A small, covered pail or bucket is an ideal place to temporarily dump food scraps for composting. Place it under the sink to keep it out of the way until you are ready to carry it outside to your compost bin.

Will you need a contractor?

Recycling and composting can be done immediately and require nothing except the desire to do it. Contact your local trash pickup company and request a free recycling bin (you may also be able to get a free compost bin). While not every town recycles, many do and will have specific rules for how to separate items.

What are the best sources for materials?

For recycling, all you need is a place to stow your empty plastic, aluminum, and glass containers. To make a compost bin, locate a 3-ft. × 3-ft. area in your yard that is not in direct sunlight. It shouldn't be so large you cannot reach the center. It has to be on bare dirt or grass and not on a deck or patio.

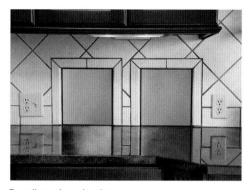

Recycling only works when you provide a convenient area to store it. Integrated recycling chutes are an elegant method of integrating a recycling center into your kitchen.

Select a spot that is convenient for you to carry your scraps to. You can start a compost pile any time of year, but many people like to start in the fall and compost their leaves.

A compost bin can be made out of four simple garden stakes and chicken wire, or it can be more enclosed with pieces of scrap wood and fencing. You can even take a trash can and punch holes into the sides and bottom. Any enclosed area is fine as long as air and soil can get in. Premade compost bins are available from gardening stores or online sources.

Once the compost bin is built, fill it up with yard trimmings (grass, branches, leaves, sawdust) and nondairy and nonmeat food scraps (vegetable peelings, coffee grounds, fruit skins, egg shells, etc.). The compost pile should be damp (like a wet towel) and may need occasional watering. Use a pitchfork or shovel to churn the pile once a month (except in the winter). Eventually, the pile will decompose into a hearty, natural mulch. The compost is ready for use when it transforms into a dark brown soil with a nice, earthy smell. None of the original "ingredients" should still be visible.

You might be wondering if a compost pile smells like a trash heap. Don't worry. The pile should not smell at all. If you smell an odor, use less water and churn the pile again. Air prevents odor. Adding more dry leaves will also help.

How much maintenance will be required after installation?

Recycling can become a simple household chore just like taking out the trash. If your building or neighborhood doesn't recycle, call the local company and demand it.

A compost bin needs to be kept damp and fluffed periodically. The compost pile will feel warm to the touch, as the decomposing process gives off a little heat.

How long will the project take to accomplish?

Tossing an empty can into a recycling bin is all it takes to recycle. You can construct a composting bin in just a matter of hours, then start collecting your kitchen and yard scraps and toss them in the bin. A premade compost bin comes preassembled and can be used immediately.

A homemade compost bin will turn your food scraps into a healthy mulch for your yard.

KEVIN SAYS It is so easy to recycle these days. Even Manhattan—the most citified place on the planet—is committed to recycling, fining residents for not separating their trash into appropriate bags. Composting is certainly the next step in solving the trash problem but offers a few more challenges.

First, you need space for a compost bin. Luckily not too much space is required, but the smaller the yard, the fewer trimmings to add to the pile. Composting also requires attention and work, and that may not be part of your regular routine.

What is the capital cost? $$

The cost of recycling is minimal—in fact, in many communities recycling bins are provided to residents free of charge. If you've got to pay for a bin yourself, then a nice can is about $50. For composting, you can spend anywhere from $30 and some weekend labor for a homemade container, or you can spend a couple of hundred bucks for something fancy. All in all, less than $300 will get you recycling and composting.

RESOURCES

Real Goods
www.realgoods.com

What financial resources are available? $$

Since municipalities stand to gain the most economically from recycling and composting, they actually are the best resource for savings. If your city subscribes to www.recyclebank.com, you can get points for recycling that you can trade in for coupons at major stores. Local government programs in California, Massachusetts, and Pennsylvania provide rebates of $25 to $60 to residents who compost.

What is the monthly cost or savings? $

Recycling could save a little money if you are creating so much recyclable trash that you can reduce the number of bins you need. Of course, this assumes your trash service charges by the bin. While this definitely saves energy for the planet, the only composting consumers who will see real savings are the gardeners who were previously buying compost. If you use compost every year, you are paying roughly $25 to $50 per cubic yard, depending upon whether you buy in bags or have it delivered in bulk from nurseries. A midsize yard might use as much as 8 yd. to 10 yd. Amortized over 12 months, you could save as much as $42 a month.

What is the long-term home value? $

There is a certain attraction to homes with great gardens that have a compost area in a discreet location. There is no data, however, to support an increase in value solely for the composting setup itself. For those looking to compost, a home with a compost bin already in place might be a convincing lure over another less environmentally friendly home.

THE BOTTOM LINE is...

Recycling is a must. It makes you feel good and there is zero downside. Okay, so you won't save money, but you don't lose money, either. And if the cash is your first priority, you can always take your cans and bottles to the recycling center and pick up a few pennies. As for composting, it's only a little hassle, and even if you can't use all the compost you make, maybe you can share it or even sell it to your neighbors. ▉

21 green home projects you can do tomorrow

If you're planning to remodel or update your home, these projects will allow you to greatly reduce energy and water use. Plus, nearly all the projects will improve the health and livability of your home without sacrificing appearance or comfort.

Add solar power

Green $pecs

Overall Rating

Difficulty

Green Benefits

ERIC SAYS Buildings are the largest single users of energy—the lights, television, and other appliances and gadgets we have take a toll—and keeping them running consumes most of the world's energy. But what if we could fuel our homes with a free and unlimited source of energy? That is the potential of putting solar panels on your home.

What will this project do for your home?

Depending on where you live, solar panels can offer a unique opportunity to free yourself from a monthly electric bill, or at least a sizable portion of it. Results will vary from state to state depending upon how much sunlight you get, but with no moving parts, no emissions, and virtually no maintenance, solar panels can provide clean, worry-free energy to your home for decades.

What will this project do for the Earth?

More than half of the electricity in the United States is produced by burning coal. Unfortunately, coal is a fossil fuel, left over from the ancient remains of dinosaurs and plants. When we run out of coal, we cannot produce more in less than, say, 150 million years. The destructive effects of coal mining and air pollution make coal the single largest polluting industry in the United States. If that wasn't enough, coal is the single greatest source of greenhouse gas emissions.

By producing your own power with solar panels, you are reducing our dependency on fossil fuels like coal. In fact, adding solar panels to your home is perhaps the most effective thing you can do to combat global warming.

Will you need a contractor?

Solar panels are typically installed by local solar providers. If you're working with an architect or contractor on a remodel, they should also be able to provide the names of local solar providers. Ask for references and make certain they are both licensed and insured.

No solar installer will be able to give you an accurate estimate without first visiting your home and looking at the roof. During the visit, the installer will look at the amount of sunlight on the roof, the possible methods of attaching the panels, and the connection to the electric meter. By the end of the visit, the installer should be able to give you a strong sense of the possibilities. A firm estimate should arrive a few days later.

In a grid-tie system, solar panels feed energy into the neighborhood electrical grid all day, causing your electric meter to run backward. At night, your home pulls energy off the grid, and the electric meter runs normally. The digital meter indicates whether energy is going back into the grid or being pulled off it.

What are the best sources for materials?

Panels are available from numerous sources online, including Amazon.com and SolarHome.org. If you feel at all intimidated by do-it-yourself projects, think no more and hire a professional.

Solar panels are typically placed on a roof (and can be installed on nearly any type of roof material), but they can be placed elsewhere. Any unobstructed and sunny spot within 1,000 ft. of your home works just as well. The height and unobstructed view of the roof makes it a preferred choice for solar installations. Several types of panels exist:

- **Flat panel collectors:** This is the typical solar panel you've probably seen before, in a long, rectangular shape and only a few inches thick.
- **High-efficiency flat panel collectors:** A variation on the typical panel, these panels look the same but produce slightly more power and cost more. They are used in locations short on roof space.
- **Flush shingle style:** Shaped like roofing shingles, these panels are installed in line with a home's regular roofing. In fact, they are the roofing. They are best used in neighborhoods with strict design requirements, or if you have a pesky neighbor complaining about the appearance of solar panels. Although they cost about 20% more than standard panels, these solar shingles replace the roof, saving you a little bit on the cost of standard roofing material. Since they blend seamlessly with your roof, most neighbors (and design review boards) won't even know they're there.
- **Solar concentrators:** This is a new technology that uses mirrors to focus sunlight onto a smaller solar panel. The high concentration of light allows the panel to produce more energy per sq. ft. than panels

alone. While these are just now becoming commercially available, expect to see them in widespread use in the next few years.

You will need at least 200 sq. ft. of south-facing, unobstructed, un-shaded space on the roof or in the yard to fit your solar installation. Most panels work best when angled roughly 10 to 40 degrees. Most roofs are not at the appropriate tilt angle, so panels are placed on frames to raise them to the correct tilt. A flat roof is an ideal location as the panels can be turned and angled as desired without being as visible from the street.

Since solar panels require sunlight to create electricity, they only work in the daytime. In the old days (10 to 15 years ago), large banks of batteries would be installed in your basement to store the power. The solar panels would charge the batteries during the day and the house would drain the batteries at night. Today, batteries are rarely used, switching instead to what is called a "grid-tie" system.

Grid-Tie System

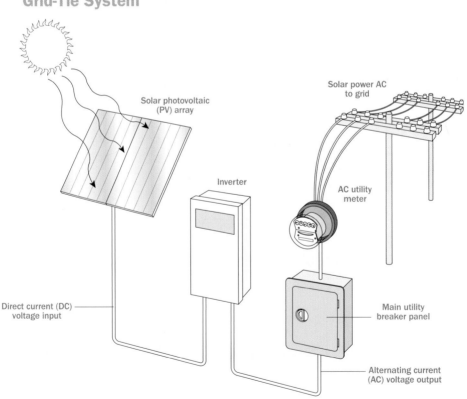

Solar photovoltaic (PV) array

Solar power AC to grid

Inverter

AC utility meter

Direct current (DC) voltage input

Main utility breaker panel

Alternating current (AC) voltage output

With a grid-tie system, your house remains connected to the public utility service. During the day, the panels produce energy. Any unused electricity is fed back into the electrical grid and your electric meter runs backward. At night, when the sun is down and the panels are inactive, your home is still connected to the utility grid. You simply draw the power you need and your electric meter runs forward. The net result can be a monthly utility bill of $0, and in some cases the utility companies will even purchase excess power from you.

The number of panels your house will require to reach that zero balance at the end of the month is estimated by your installer, but it is not an exact science. Before installation, your solar company will want to review your past electrical bills to get a good idea of the amount of energy you typically use. They use this information to size your solar system. You don't want too large of a system, or you would produce more power than you need. Too small a system will leave you still buying power from the local utility company.

If cost is an issue, or if you can't afford to install solar panels to meet 100% of your power needs, don't panic. You can start off small, with fewer panels, and use a combination of solar power and grid power. The panels are wired together in a series, so it's simple to add more. But do yourself a favor: Install all of the mounting brackets on your roof at one time. It will save you money when you decide to add more panels and will reduce the number of holes drilled into your roof.

If you are planning to replace your roof in the next five years, it would be best to do at the same time as the solar installation. Having the mounting brackets installed below the roofing greatly reduces the chance of roof leaks. If you can't afford solar at all right now, then the next time you replace the roof, install an electrical conduit. It will make things much easier for when it comes time to install a solar panel system.

Solar energy is not just for the West Coast. It is the light, not the heat, that allows a solar panel to generate electricity. A cold and sunny location works just as well as a warm and sunny location. In fact, the panels continue to produce power on cloudy and foggy days, just not as much.

How much maintenance will be required after installation?

Solar panels require little maintenance other than an occasional cleaning to remove any grime or dust. Of course, in areas with seasonal weather you will have to regularly clear off leaves, snow, ice, and any other debris. The panels produce more energy when clean.

If you are moving to a new home, you can take the panels. It is not like wall-to-wall carpeting. Think of the solar panels as an appliance. With a little effort and preparation, they can be moved to a new location.

How long will the project take to accomplish?

Once you have selected your installer, the process goes quickly. Large installers usually carry panels in stock, while others may have to order the panels, which might take a month or so. When the parts are available and the installer begins work, the typical residential roof installation takes about one to five days, depending on the size of the solar panel system and access to the roof.

KEVIN SAYS Solar power is an emotional green win for most people. We like seeing tangible savings in a regular monthly electricity bill. It makes us feel like we're taking back control from those big monopolistic energy companies and government agencies. For most people, the biggest negative about installing solar power is figuring out where to put the panels. To have an impact on the home, you need a fair amount of space, and the roof of your Cape Cod home wasn't designed to hide giant dark blobs of silicon attached to pipes and wiring. For years, builders avoided installing solar systems because neighbors and buyers alike weren't willing to sacrifice the appearance of their houses and community for a few hundred dollars in savings. But these days, fashionable people are pushing hard to make "green" the new "black," so maybe showing off your panels will soon become the equivalent of putting a Warhol on your roof.

What is the capital cost? $$$$

Solar panel projects start at around $7,500 for the smallest system and can go as high as $65,000 for a larger house. According to several solar manufacturers, a typical system for a 2,500-sq.-ft. home should be about $35,000, depending upon features and geographic location. As green projects go, solar is a fairly high-ticket item.

What financial resources are available? $$$$$

Here's where solar add-ons have huge benefits. Perhaps because they have been around for so long, there are lots of programs available to help you pay for solar. The cost is high enough that refinancing your home or pulling funds off a home equity line can be a feasible way to finance the project. Additionally, almost every state government has some sort of solar energy program comprised of grants, low-interest loans, and tax credits (details vary with location). Visit the Database of State Incentives for Renewable Energy (DSIRE) website at www.dsireusa.org and find plenty of ways to get back some of those state taxes you have been paying while financing your solar energy panels at the same time.

What is the monthly cost or savings? $$$

The good news is that solar definitely saves money on a monthly basis since you use your own power instead of the utility company's energy. The exact savings vary depending upon several factors including where you live, the size of the house, and the amount of energy used. If you live in a small house where there is not much sun (think Seattle) and you pay less than $100 monthly for your electricity, your savings are likely to be fairly minimal. Assuming a 50% savings on energy, even a $50 monthly savings would take you more than 12 years to recoup the cost of the smallest solar system. If you are able to use credits to abate the cost, you might gain ground in a shorter period of time, but if you finance the project, the interest will add to the time needed to pay for the system.

On the other hand, for big-home, sun-city dwellers who run the A/C constantly, solar should pay off in shorter time. If you assume a net capital cost (after credits) of $25,000, and a $250 monthly savings, you can make your money back after just 8 years. Either way, once you make it past the break-even on a paid-off system, you are in the green both ways. But solar won't usually make you completely energy green. Unless you are on a grid-tie system with an abundance of light, you can't rely solely on a solar system for all your power needs. You'll still have to pay the utility bill every month. However, a large focus of corporate research and development money and energy is going into improving solar technology and being public utility-free may be an option for many people within 10 years.

What is the long-term home value? $$

Solar companies like to claim that installing solar panels will automatically add value to your home equal or greater to the cost of the
purchase. Many quote a questionable 1998 study from the Appraisal
Institute stating that annual energy bill savings of $1 result in a $20 value increase for your home, which means a $100 monthly savings would
equal a value increase of $24,000.

I surveyed five certified appraisers in various parts of the U.S. All five
independently claimed that there is still insufficient sales data to support such claims and that lenders and lending agencies—while
accounting for monthly savings when qualifying borrowers—will not
yet allow appraisers to adjust property values upward for solar installation. Not only that, but some appraisers commented that poorly
installed systems that are plainly visible could actually detract from a
property's resale value.

Photovoltaic technology is still in its infancy. Silicon Valley companies famous for building microchips are now investing billions of
dollars in advancing our capacity for solar energy. It's likely that solar
capacity will soon be doubling at the same pace as computer memory,
which means many of the systems installed today could be obsolete in
less than 10 years.

So, let's say a buyer considers purchasing your home (which has
solar panels), and compares it to a similar home across the street
(which does not have solar panels). Even if the buyer found your home
to be more desirable due to its solar panel upgrades, they would likely
account for the advances in technology and weigh the additional asking price of your home with its older-technology panels against the
cost of installing a brand-new system on the other house. Since the
buyer would get all of the benefits of solar energy credits and tax savings on the new system, your system would most likely be perceived as
less valuable to the buyer and would only add a fraction of its original
cost to the value of your home.

THE BOTTOM LINE IS...

Big energy users in light and bright areas of the country can make solar
pay for itself in a reasonable time frame. However, if you plan on selling
the house soon or don't use more than $150 monthly in electricity—and
aesthetics are critical in your neighborhood—you may want to wait for
the price reductions coming in next-generation solar technology. ▥

Install
micro–hydropower

ERIC SAYS If you have running water on your property, it is more than a beautiful natural wonder. Your stream could provide a clean, limitless supply of energy for your home. The concept behind hydro (as in water) power is really just a conversion of energy. The natural flow of a river or stream is used to spin a turbine. That turbine then produces electricity. Although the components of the system can get complex, the basis of hydropower is simply capturing the kinetic energy already within moving water. Dams are examples of hydropower done on a large scale, but micro–hydro (as in small) systems can be sized for the more modest needs of a home.

Of course, to use a micro–hydro system, your property needs to have flowing water, and most properties do not. But if you live alongside a river or have a babbling brook running through the site, a small and efficient micro–hydro system is one of the most renewable and reliable sources of energy for your home. As long as the water flows, so will the free energy.

What will this project do for your home?

If you have water on your site, micro–hydropower could be an attractive option to get rid of your monthly energy bill forever. Micro–hydropower also offers a few advantages over other renewable energy sources. Unlike wind turbines, micro–hydropower is a quiet source of energy production. And unlike solar energy (see p. 90), micro–hydro is a 24-hour energy source.

Green $pecs

Overall Rating

Difficulty

Green Benefits

Energy can be obtained from flowing water, no matter how small the stream. If the stream is often filled with debris (leaves, twigs, silt), then an additional filter might be needed. If the stream freezes in the winter, the micro–hydro system must be drained and put away before the first frost.

What will this project do for the Earth?

The majority of the energy consumed in the United States comes from dwindling and polluting sources. By no longer relying on coal or natural gas for energy, you directly help stop the effects of global warming. Plus, by making the power right at home, you remove your impact on the local utility system. And micro–hydropower does not pollute or consume any water. Any water used is pumped right back into the stream.

Will you need a contractor?

Installing your own micro–hydro system is popular with the do-it-yourself crowd but not something I'd recommend for most of us. The system is complex, with many components to connect. You can find networks of DIY hobbyists discussing their experiences installing and maintaining such systems online, but unless you're comfortable with the idea, get a professional installer.

What are the best sources for materials?

Commercial micro–hydro systems and components are usually sold as a package and can be found online or through electrical catalogs. If you do attempt to do it yourself, the systems require you to select the proper generator with the correct turbine. Be sure to ask the manufacturer for help if you're unsure about how to proceed.

In remote or rural locations, micro–hydro systems are often connected to a backup bank of batteries. Since you can't possibly use all of the energy at the exact time it is produced, batteries store excess power for times of slow water flow or heavy electric usage. Solar panels only produce energy during daylight hours, so their batteries must be large enough to provide power through the night. The consistent flow of water provides an equally consistent charge to batteries, allowing for a much smaller battery bank. Technically, if your stream has enough potential energy, you could skip the batteries entirely.

If your home already has public utility service, a grid-tie system might be the answer. Any surplus energy you create is pumped back into the public energy grid, forcing your electric meter to roll backwards. Many public utilities accept electricity generated by solar panels, but they might not be as receptive to a micro–hydro system. Check with your local utility for restrictions. Their answers may help you decide whether to use grid-tie or battery backups. If you do use a battery bank, it should be located as close to the turbine as possible to avoid energy losses from transmission.

In addition to utility requirements, local permits and water rights are other issues to research before embarking. Even if you have water access rights, you may need a separate water right to produce energy. Your first call should be to the local county engineer's office. In addition, you'll need to contact the Federal Energy Regulatory Commission (http://www.ferc.gov, 866-208-3372) and the U.S. Army Corps of Engineers (http://www.usace.army.mil, 202-761-0011). Permits in some jurisdictions can take up to a year to obtain.

How much maintenance will be required after installation?

Maintenance is minimal. Periodically checking the piping for leaks or caught debris is necessary but easily done. A micro–hydro system is sometimes more reliable than the local utility and will last a minimum of 20 years if properly maintained.

How long will the project take to accomplish?

Installation of a micro–hydro system varies based on size but can be done in a couple of weeks. During installation, the location of the piping should be coordinated with any other major grading or work planned at the site.

KEVIN SAYS Micro–hydropower is a great idea in theory, but it doesn't apply to everyone. Not only do you need to have running water on your property, but you also need to have the legal rights to use that water. Luckily, most micro–hydropower systems don't negatively impact the flow, but there can still be a fair amount of red tape. Then there is the maintenance. This is not a flip-a-switch-and-go type of system. It requires periodic cleaning and monitoring, as well as an understanding of how the systems work.

Many of us are thrilled with the idea of generating an abundance of power, which brings up the issue of selling excess power back to the utilities. This idea gets many people excited until they remember that they have to be connected to the grid in the first place. For most homeowners who investigate micro–hydropower, the real obstacle is bringing the grid to their remote property. If your property is even just 1,500 ft. from the grid, it can cost you more than $20,000 to connect. If you are a mile away, you could be looking at more than $100,000, which makes the idea of selling back power useless.

What is the capital cost? $$$$

The total cost for micro–hydropower depends on how much power you will need. Complete systems start at around $5,000 but can go easily to $20,000 or more (up to $100,000 for an industrial-grade system) for a relatively small power yield.

What financial resources are available? $$

If you are lucky enough to be situated with running water and can tie into the grid, you may be able to get tax benefits if you live in North Carolina or Vermont. Otherwise, there is not much in the way of publicized programs specifically for micro–hydropower. That being said, several states and municipalities have programs for solar power that are broadly worded to include other alternative power sources, so check with your local agencies. The federal government has focused mainly on wind and solar and left water power out of the mix.

Micro–Hydro System

A micro–hydro system starts upstream, where a filtered intake diverts water into a pipe called a penstock. The water flows in the pipe, fueled by gravity and the pressure of the stream. Downstream, the water flows into an electrical turbine, spinning the turbine and creating electricity, before it is discharged back into the stream.

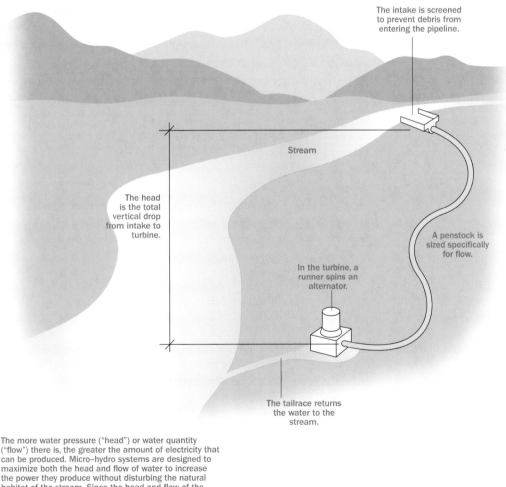

The intake is screened to prevent debris from entering the pipeline.

Stream

The head is the total vertical drop from intake to turbine.

A penstock is sized specifically for flow.

In the turbine, a runner spins an alternator.

The tailrace returns the water to the stream.

The more water pressure ("head") or water quantity ("flow") there is, the greater the amount of electricity that can be produced. Micro–hydro systems are designed to maximize both the head and flow of water to increase the power they produce without disturbing the natural habitat of the stream. Since the head and flow of the water determine the energy output, hilly sites tend to work best (but not always). There needs to be a drop in elevation of at least a few feet. Unlike dams, which interrupt the entire flow of a stream or river, micro–hydro is barely noticeable once installed.

What is the monthly cost or savings? $$$

Savings from micro–hydropower occur if you are already on the grid or able to connect to one. Otherwise, there's nothing to save. If you are on the grid, then you are likely saving a couple of hundred dollars each month for the energy provided by a small hydro system. At a $200 monthly savings, you would need at least a couple of years before the smallest system would pay for itself, but you are home free from then on.

What is the long-term home value? $$

A home that is off the grid already commands substantially less value in the marketplace than a home connected to the grid. Although the general nature of lower demand for rural property is a factor, the main reason for lower values is lack of financing. Federally owned Fannie Mae and Freddie Mac—the organizations that buy most mortgages—will not buy loans on properties off the grid, so most banks follow suit even if the homes have reliable power systems. This means only very wealthy people can afford to purchase these properties, and few of them seem to want homes in remote areas. Simple supply and demand keeps these values down. If the property has a reliable power system, it can command a substantially higher sales price than a rural property with no power at all (perhaps more than the cost of the system itself).

THE BOTTOM LINE IS...

If you are lucky enough to be on the grid and have a legally accessible water source, you can probably get bang for your buck by adding micro–hydropower. You won't see a ton of savings, but it can make for a steady supplement to your power needs. For those who don't have access to a power grid source, micro–hydropower may be one of the few ways to make the house a viable living space and give it worthy resale value. ■

Bring in natural daylighting

ERIC SAYS Before the widespread use of electric lighting changed everything, architects and builders constructed buildings to take advantage of natural light. The arrangement of spaces, the depth of rooms, and the height of hallways were determined by a need to bring natural daylight deep into a building. Even after electric lighting became commonplace, older buildings remained inherently energy efficient because there was no need for extra lighting during the daytime. But after a century of lighting technology and (what used to be) inexpensive energy, architects and builders have all about lost the knowledge of how to use natural daylighting.

The average home uses a quarter of its electricity for lighting. Daylighting—the use of natural light to illuminate spaces—could save nearly half of that energy. Using natural daylighting is the easiest way to take advantage of solar energy in your home, but it does not mean letting in as much natural light as possible; it means bringing in a controlled amount of light.

There are several strategies to incorporate daylighting into your home beyond just using windows. In a minor home remodel, you can brighten up dark rooms with the careful addition of clerestories or skylights. Using clerestories or skylights brings light into the home without giving up privacy and makes the most of spaces where windows are not possible.

What will this project do for your home?

We flourish in sunlight. Exposure to sunlight aids in relaxation and happiness and can help reduce seasonal affective disorder (SAD), also known as the winter blues. Natural daylighting has also been proven to increase student test scores in schools, double retail sales in stores,

Green $pecs

Overall Rating

Difficulty

Green Benefits

and boost worker productivity in factories. Daylighting can also cut lighting bills by half.

What will this project do for the Earth?

By reducing electric lighting use in homes by 25%, natural daylighting could save enough energy to power 17 million homes per year. That entire energy savings also translates into a drop in carbon emissions and global warming.

Will you need a contractor?

Installing clerestories and skylights into an existing wall or roof is best left to a licensed, professional contractor. A poorly installed clerestory or skylight can lead to air drafts, wall cracks, and water leaks.

What are the best sources for materials?

In looking at adding daylighting to your home, it is important to remember two things:

1 Not all clerestories and skylights are created equal. Thermal systems are worth the higher cost in severe climates.
2 More light is not better light. You want an even, controlled amount of light in each room.

Clerestories A clerestory is a window located high on a wall. Clerestories bring in natural light but maintain privacy. They are great

Skylight Specs

Essentially a fancy type of horizontal window, a skylight lets in an incredible amount of natural light and fresh air and also helps push noxious gases and odors out of the home. Light brings heat too, so place skylights carefully to avoid overheating.

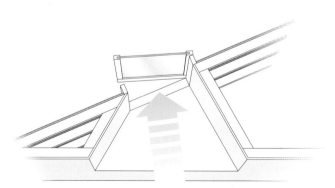

Tube skylights install through existing attics, making them ideal for remodeling projects. They can bring daylight into once dark areas of a home.

for bathrooms, rooms that face a public street, or rooms where wall space is at a premium. Place a clerestory as close to the ceiling as possible. The light will bounce off the ceiling and bring a warm glow to the entire room. The ceiling should be a light color to aid in this reflection but does not have to be white. If the clerestories are used for natural ventilation, an operable lower window on the opposite wall will help fresh air circulate throughout the room and not just hover at ceiling level.

Skylights Skylights are really a fancy type of horizontal window. Because of their orientation, they can allow an incredible amount of light into a room. One standard-size skylight is enough to brighten an entire room. But remember, all of that light will bring with it a lot of heat. Carefully consider the locations of skylights. If you live in a hot place, a skylight may not be the wisest choice for your home.

To help control this increase in heat, manufacturers often offer tinting, retractable shades, and angled tops that face away from the sun. Operable skylights are also available. Whether installing skylights in a new home or in your existing one, keep in mind that skylights are prone to leaks. Only choose a self-flashing model and insist that your contractor follow the manufacturer's recommended installation procedure.

Tube skylights Skylights can provide a significant increase in the amount of light within your home, but for them to work there has to be a roof directly overhead. If you have an attic covering your home, a tube skylight (often referred to as a sun tube, sun tunnel, or sun pipe) could be the solution.

Residential electricity is
31% of all electricity use,
costing $37 billion annually.

$37 billion
× 31% residential
electricity use
= $11,470,000,000
residential electricity use

Lighting is 25% of
residential electricity use
(and goes up to 60% in
commercial spaces).

25% of $11.47 billion
= $2,867,500,000 that
could be saved with natural
daylighting

According to research by
the Department of Energy,
national residential electricity
consumption in 2004 was
1.29 billion megawatt hours
(mWh).

25% of 1.29 billion
= 322,500,000 mWh
= the output of 90 power
plants

A small dome is installed on your roof, similar to a standard skylight. Inside the attic, a flexible duct connects the dome to the ceiling of the room being lit. Sunlight enters the dome on the roof, bounces around inside the duct, and travels down to a round lens.

Tube skylights are much less expensive to purchase and install than standard skylights. Their simple design reduces the chance of leaks, and the reflected light does not bring in as much heat as a standard skylight. Since they connect to the roof through a flexible hose, tube skylights can be snaked through an attic or wall to drop light into unexpected locations. You could even install them in a basement and run the tube through a closet on the upper floor! And most manufacturers offer lighting trim kits so the lens can provide lighting at night.

Daylighting strategies A frequent mistake is to assume the more daylighting the better. Instead, you want controlled light in the correct places. Keep in mind that sunlight can only reach a certain distance into a room. As a general rule, light will penetrate two-and-a-half times the ceiling height into a room. For example, a room with a 10-ft.-tall ceiling will allow the sun to reach 25 ft. into the room during the middle of the day. The ideal daylighting solution admits light but not summer heat; admits light and heat in the winter; allows views but not glare; reflects off walls and ceilings to provide an even glow; and combines with dimmers or daylight sensors when electric light is needed.

The placement of clerestories and skylights has direct impact on what kind of light, heat, and convenience they will provide. North-facing clerestories and skylights get no direct light but provide an even glow from reflected light all day. In hot climates, they have almost no heat gain. In cold climates, a north window will be cold and gray. Those facing east will receive sunlight in the morning and start warming up the house at the beginning of the day. The angle of the light will create glare and will not penetrate deep into the house. West-facing windows receive hot afternoon sun until sunset. The western sun is much lower in the sky, so overhangs will not prevent heat and glare from entering the home. East- and west-facing windows provide glaring light and their use should be limited. The low angle of the sun at sunrise and sunset allows sunlight to penetrate deep into a home at times, creating glare that overhangs won't fix.

Since south-facing windows provide sunlight throughout the day, they are unique and a critical part of any daylighting strategy. They receive sunlight nearly all day. In hot climates, use overhangs above the windows to block the summer sun. A 2-ft. overhang would shade against the summer sun but allow the winter sun to come in, as you can see in the drawing at right. Plus, if you install a floor made out of stone, tile, brick, or concrete, the winter sun will warm up the floor, which will remain warm into the night. A high-mass floor will also absorb the heat from the sun and keep the home warm.

In a place like Arizona or Florida, try to avoid large, exposed clerestories or skylights along the south- or west-facing walls of your home. East- and north-facing windows will provide light but won't get too warm. In places with cold winters, like Maine or North Dakota, use the sun to warm up the house by placing windows on the south- and west-facing walls.

If you live in a place with:

- **Hot summers, hot winters (Phoenix, Arizona; Miami, Florida):** Use north-facing windows for light and views. Tall and narrow windows or short and high windows with overhangs work best on the south. Avoid skylights.
- **Hot summers, cold winters (Philadelphia, Pennsylvania; Chicago, Illinois):** Large south-facing windows with overhangs will work year-round. Small windows are best to the north.
- **Mild summers, mild winters (Portland, Oregon; San Francisco, California):** Large windows are fine, but they should be operable to allow for cross ventilation when needed. Skylights are ideal.

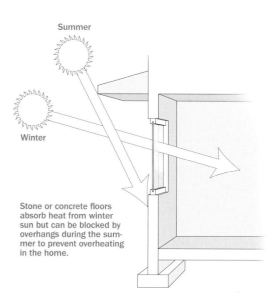

Overhangs in Winter and Summer

Overhangs can shade the higher summer sun but still allow the low winter sun to come in and warm up a home.

Summer

Winter

Stone or concrete floors absorb heat from winter sun but can be blocked by overhangs during the summer to prevent overheating in the home.

How much maintenance will be required after installation?

Maintaining skylights and clerestories is just like maintaining regular windows in the home. Check them annually for cracks around the frame and areas that need to be sealed to prevent drafts.

Room Depth Affects Lighting

No matter what size room you have, natural light only enters so far. The sun will penetrate two-and-a-half times the ceiling height (A) into any room.

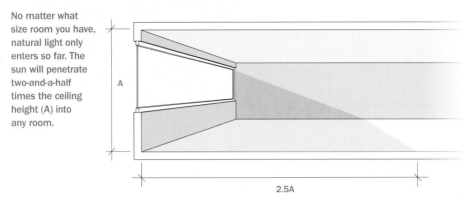

A

2.5A

Clerestories Give Light and Privacy

A clerestory is really just a high window that permits light to enter a room but maintains privacy. The trick to installing clerestories is to push them as close to the ceiling as possible to avoid dark areas.

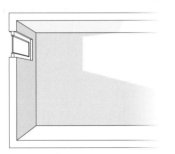

Clerestory not flush with ceiling

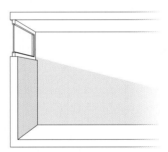

Clerestory flush with ceiling

How long will the project take to accomplish?

Replacing existing windows and clerestories can take several weeks, depending on the number of windows, number of floors, and type of exterior siding on the home. Skylights and sun tunnels can be added to a new or existing home in a couple of weeks. Get estimates for cost and schedule before hiring a contractor. It will likely take a couple of months for the windows to arrive after being ordered. Plan ahead to avoid getting caught midway through an installation in the middle of the winter or rainy season.

KEVIN SAYS Few of us want to live in a dark and dingy cave. If your home is dark during the day and you spend a lot of time inside, then adding daylighting is worth considering for reasons beyond the energy factor. We need daylight to keep us happy and healthy. Skylights have always been popular, but as long as there have been skylights, there have been complaints about leaks. The jury is still out on the tubular skylights found in the big-box stores. The concept of flexible, directed lighting sounds good, but it takes a skilled designer to place them where they will have positive aesthetic beauty, and there is plenty of Internet feedback about tube skylighting gone wrong. Clerestories done right can add great design impact to the home and serve the energy cause at the same time.

What is the capital cost? $$

The cost of daylighting depends on two factors. First, how much lighting you need, and second, the cost of installation labor in your area. Skylights range in price from $150 to $350 for standard skylights, although you could pay as much as $1,000 for a motorized version, which could defeat your energy savings. Tube lights run $175 to $500 each. Thermal windows can be used for clerestories and range from $70 to $250 each, depending on size.

Your installation costs will generally be lower if you have a lot of work done at the same time since a crew will already be out there to cut, install, patch, and paint. Your contractor may also be able to get a better price for you from their wholesale suppliers, providing they don't mark up the materials too much.

Let's say you have a midsize home of 2,500 sq. ft. and decide to add two skylights, four tube lights, and four clerestories. Your total equipment cost is roughly $2,200 to $2,500. The additional installation costs—including other materials and permitting—will likely be another $2,000 to $5,000 for a worst-case total cost of $7,500.

What financial resources are available? $$

New federal regulations allow for a 30% tax credit of purchase costs for skylights and windows up to $1,500. But the units must have a U-factor and solar heat gain coefficient (SHGC) less than or equal to 0.30, which shows on the National Fenestration Rating Council (NFRC) label. Some

DO THE MATH

In 2006, the average monthly residential electricity consumption was 920 kWh. We use 25% (or 322,500,000 mWh, see p. 106) of that electricity for lighting our homes.

322,500,000 mWh / 920 kWh (9.2 mWh) = 35,054,347.826 homes that could be powered on the energy we use just for lighting

Daylighting could save up to half of that electricity and power 17,527,713 homes.

RESOURCES

Natural Lighting Company
www.daylighting.com

states such as Arizona and cities like Pasadena offer additional rebates of a few hundred dollars as well, so it's worth checking around.

What is the monthly cost or savings? $

The primary money savings for daylighting comes from reducing the amount of electric lighting used in the house. The only time you save money is during daylight hours when you are at home. If you work a 9-to-5 schedule, with a reasonable commute, you may only be at home an average of three to five daylight hours per day, which represents roughly 20% to 35% of your total usage. If you are typically home during the day, you might see larger savings.

You might reduce a $200 energy bill by about $10 to $20 per month if you were previously a heavy daytime electric light user. In the best case, recouping the installation cost on monthly savings would take you roughly 40 years, which isn't very exciting. Additionally, servicing the flashing on skylights may require maintenance expenditures—particularly in wet climates—negating any savings and possibly adding cost.

What is the long-term home value? $$$

No question that a light-and-bright home outsells a dark home every time. While bank appraisers don't give a specific value to homes that are light and airy, they will give relative value in design if the home takes advantage of natural lighting. Real estate agents typically have sellers paint dark walls white and remove dark shades for that purpose. Prominent skylights are considered welcome additions that will generally recoup their cost in sales value as long as they provide beneficial light and don't leak.

THE BOTTOM LINE is...

For a new home, daylighting design makes all the sense in the world. On a remodel basis, the advantages are debatable. If you are a heavy daytime home dweller who needs lots of light, you might see a little financial benefit from adding daylighting. Certainly a dark, cavelike home is worthy of opening up to the sky, but don't look to daylighting as a necessity in most new homes or as a strong benefit to your wallet. ▦

Replace your windows

ERIC SAYS Windows bring natural light, fresh air, and views into your home, but you pay a price for this transparency. As much as 40% of a home's heat and air-conditioning is lost through windows. In addition, that extra sunlight can sometimes bring in extra heat. Window technologies have advanced so much in recent years that new windows are twice as energy efficient as those from just 10 years ago, so switching out old windows with energy-efficient replacements may be well worth it.

Green $pecs

Overall Rating

Difficulty

Green Benefits

What will this project do for your home?

Depending on the condition of your old windows, replacing them with new, energy-efficient sashes can reduce your heating and air-conditioning bill by up to 30%. Installing double-pane windows reduces heat loss (heated air leaking outside) by half compared with those old-fashioned single-pane windows. The efficiency of high-performance windows also allows you to have large, expansive areas of glass with little effect on energy usage.

What will this project do for the Earth?

If every home in the United States replaced their old, leaky windows, it would conserve enough energy to heat and cool 2.2 million homes per year. That is the equivalent of taking more than 323,000 cars off the road.

Will you need a contractor?

Removing old windows and waterproofing and installing new ones is not for a beginner. Buying the most energy-efficient windows available is only worth it if they are installed correctly, so I suggest hiring an experienced and licensed contractor to do the job.

According to www. efficientwindows.org, 27% to 39% of a home's heating bill could be saved with energy-efficient windows. We'll go with 30% for this estimate.

There are 111,000,000 households in U.S., and 82 billion kW are used by homes each year for heating and cooling.

30% savings = 24.6 billion kW

In 2006, the average annual residential electricity consumption was 11,040 kWh, which is enough to heat and cool 2.2 million homes:

24.6 billion kW / 11,040 kWh = 2.23 million homes

24.6 billion kW × 12.4 cents per kW = $3,050,400,000 in potential savings if all homes in the U.S. installed new, energy-efficient windows

See http://www.eia.doe. gov/emeu/recs/recs2005/ c&e/spaceheating/pdf/ tablesh1.pdf for more information.

What are the best sources for materials?

Window prices vary greatly based on the manufacturer, so request estimates ahead of time. Custom-size windows are not much more expensive than standard-size windows. Many manufacturers no longer stock standard sizes, preferring "semicustom" sizes instead. Most window manufacturers will not sell directly to the public. Ask your contractor to purchase the windows. You may be able to find local wholesalers of windows, but be careful of the quality.

The primary reason to replace windows is to improve a home's energy efficiency. This improvement in energy efficiency is based on the window frame, the type of window, and the type of glazing.

Window frames The window frame is the structure of the window. It conducts heat and cold, and contributes to the overall energy efficiency of the window. Frames generally come in four materials: wood, aluminum, fiberglass, and vinyl.

- **Wood frames:** Wood window frames cost more than any other type, but are the best insulated. Bare-wood frames must be painted (and repainted), which means a great deal of maintenance. Many manufacturers offer wood frames clad in a metal skin, so windows never need painting and last 30 years. The greenest option is to purchase clad wood frames made out of wood that has been sustainably harvested and certified by the Forest Stewardship Council (FSC). Expect to pay 20% more for the FSC certification.

- **Aluminum (or metal) frames:** Strong, lightweight, and maintenance free, metal frames seem like an ideal choice. Unfortunately, metal conducts heat and cold, and is a poor insulator. Although frames are typically made from aluminum, steel and other metals are also available. Look for metal frames with a "thermal break"— an insulating gasket separating the two halves of the frame, which improves the insulation of the window.

  Try to find metal frames with a high recycled content. Many manufacturers do not disclose (or even know) how much recycled content is in their frames, but this will change in the near future, so ask the manufacturer for the percentage of recycled metal in their frames.

- **Fiberglass frames:** Cheaper than wood and a better insulator than metal, fiberglass windows are a healthy and smart choice. Select insulated frames for the highest energy performance. Although they come in a limited number of colors, fiberglass frames never need to be painted.

- **Vinyl frames:** Vinyl is another name for polyvinyl chloride (PVC). Vinyl frames are incredibly affordable, durable, and have a good insulating value. Unfortunately, vinyl is also referred to as the "poison plastic" and the production of vinyl has been linked to serious health concerns, including a rare form of cancer. In a fire, the smoke from burning vinyl could do more harm to you than the flames themselves. In my professional advice, the low cost of vinyl doesn't outweigh its negative environmental effects.

Types of window units Modern windows are much more than a single sheet of glass. Insulated units—two panes of glass sandwiched together inside the frame—have become the standard. The air space between the glass acts as an insulator. Windows can be double– or triple–paned to increase energy efficiency, and the air space between the panes of glass can also be filled with inert gas. Usually argon or krypton gas is used, offering additional insulation. Although slightly more expensive, krypton gas provides the better insulation performance. At a minimum, select double-glazed insulated window units. Insulated units are available no matter what type of frame you choose.

Parts of an Energy-Efficient Window

Modern windows have double the energy efficiency of their older counterparts. New technologies such as double glass, gas fills, and special glass coatings vastly improve heat loss through windows.

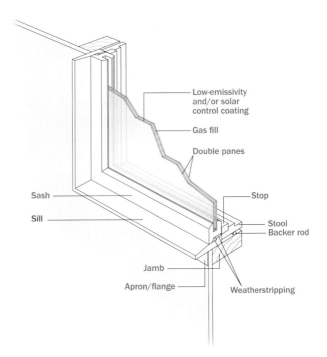

Low-emissivity and/or solar control coating

Gas fill

Double panes

Sash

Sill

Stop

Stool
Backer rod

Jamb

Apron/flange

Weatherstripping

Figuring U-Value

If all of this talk about "double-paned windows with low-e glass" is confusing, there is an easier way to figure out the energy efficiency of a window. The insulation of a window is measured by a "U-value." U-value is just like the R-value of insulation, except that the numbers are the inverse of each other. With R-values, the higher the number the better. With U-values, it is the exact opposite; the lower the number the better. Any window you purchase should have a U-value no greater than 0.30. Ask the manufacturer for the U-value of a window before buying.

Types of window glazing The bulk of the window is the glass, which is the greatest source of heat loss. There are many types of glazing or glass available, including tinting to reduce light, coatings to cut down on ultraviolet light, and low-emissivity (low-e) coatings to lessen heat gain from sunlight.

Low-e coatings control heat transfer through a window. Although they could add 10% to 15% more to the cost of a window, they can improve energy efficiency by 30% to 50%. Have low-e coating put on the outside glass if you want to keep the heat from coming in the house. If you live in a cold climate and want the sunlight to enter your home, install the low-e coating on the inside glass instead.

Select glazing based on the orientation of your house. On the south side of your home, choosing a low-e coating will let in light but block most of the heat. Installing tinted glass on the west side of the house will reduce afternoon glare. Do-it-yourself window films are also available for existing windows. They are easy to install and can add tint or low-e coatings.

Aside from looking at U-value (see the sidebar at left), a couple of certifications will also help you locate energy-efficient windows. The National Fenestration Rating Council (NFRC) tests, certifies, and labels windows based on energy performance. The NFRC label is a trusted seal that you can use to determine energy properties and to compare windows. More information can be found at www.nfrc.org. The U.S. Department of Energy's Energy Star program also certifies a minimum energy performance for appliances and windows.

How much maintenance will be required after installation?

Checking annually for drafts and leaks will ensure that windows provide the highest energy efficiency.

How long will the project take to accomplish?

If you're replacing existing windows with custom windows, it could take up to eight weeks for them to arrive from the manufacturer. I've seen more construction delays caused by waiting for windows than any other product. If the manufacturer says five to eight weeks, expect it to be the longer wait and plan your project accordingly. You don't want to be removing old windows in the winter or rainy season. For an average-size home, the process of actually removing old windows and installing new ones takes two to four weeks.

KEVIN SAYS There are several overlapping environmental issues when it comes to replacing windows. First is the idea that the waste created by replacing windows can diminish its benefit of energy savings. Then there's the aesthetics. Some argue that swapping out wood windows for vinyl can make a house look somewhat cheap. In older homes, glass has an iridescent quality that is irreplaceable and adds charm despite the energy losses. Then there is the debate over the toxic aspects of the use and production of vinyl versus its energy savings.

There is also a growing trend around saving existing wood-frame windows and simply replacing the glass and sealing them where necessary. Considering that the bulk of homes in need of window replacement are between 25 and 60 years old, the aesthetic issues may take a backseat to cost savings, and by taking the vinyl option out of the picture, it's tough to recoup replacement costs.

What is the capital cost? $$$$

At Eric's suggestion and due to the toxic vinyl argument, we will focus on wood and fiberglass window replacement options. Suffice it to say that you will pay nearly double for a wood or fiberglass window replacement over vinyl. This cost difference only impacts the cost of materials; installation cost is the same for all, although wood windows will need painting or staining. Most standard-size wood or fiberglass double-pane windows range from $150 to $300 per window. Installation can run another $200 to $250 depending upon the condition of the existing window casing.

What financial resources are available? $$

Energy Star–rated windows are a mainstay of utility rebates in many areas of the country. The amounts are generally a couple of hundred dollars at most, although Mission Valley Power in Montana has a more creative offering of $0.75 per sq. ft. If you replaced twenty 3-ft. × 5-ft.

DO THE MATH

30% of our yearly national heating and cooling energy is 24.6 billion kW (see p. 112), which can also be understood in terms of metric tons of CO_2 saved:

24.6 billion kW = 1,766,683 metric tons of CO_2

1 car emits 5.46 metric tons of CO_2 emissions per year.

1,766,683 metric tons of CO_2/ 5.46 metric tons of CO_2 per car per year = the annual greenhouse gas emissions from 323,568 passenger vehicles that can be prevented from escaping into the atmosphere

See http://www.epa.gov/cleanrgy/energy-resources/calculator.html for more details.

RESOURCES

Forest Stewardration Council (FSC)
www.fscus.org

National Fenestration Rating Council (NFRC)
www.nfrc.org

windows in your home, you would be entitled to $225. Available federal money is limited to 30% of materials up to a maximum of $1,500. All windows need to have an energy rating of lower than 0.30 to qualify, but few products are offered these days that do not meet this requirement.

There are often special rebates and discounts running at the big-box stores, but these are usually on loss-leader vinyl windows, so it's best to check with your contractor or a window specialist to compare prices.

What is the monthly cost or savings? $

Depending upon the size of your home and local utility costs, Energy Star estimates savings of $125 to $450 per year when replacing single-pane windows with modern gas-filled ones and $25 to $110 per year when swapping out double-pane windows. Best-case scenario, it takes roughly 20 years to recoup costs on the window replacement from energy savings.

What is the long-term home value? $$$

As long as new windows match the quality of décor of a home, they can add value over old single-pane windows in need of replacement. Bank appraisers will give value to new windows at a 25% to 50% discount depending upon how recently the installation took place. Buyers will definitely consider new windows as a factor in paying more for a house.

One other side benefit to new windows, especially for urban dwellers, is noise reduction. Street-facing apartments in Manhattan have increased their value significantly with the addition of double- and triple-pane windows that all but eliminate street noise.

THE BOTTOM LINE IS...

Postwar homes can gain benefits in looks, livability, and some energy savings with window replacements, but you're not likely to save tons of money. But there are options for energy savings such as coatings that can be cost-effective. Keep in mind that home ownership requires maintenance, and windows are a critical factor for making the home beautiful and livable. So when it's time, go for the best-looking and best-performing windows you can afford. ■

Install a whole-house fan

ERIC SAYS Given how outside temperatures rise and fall throughout the day, it's surprising that we are comfortable only in about a 4°F range inside. We want our homes to remain from 68°F to 72°F all of the time, yet being comfortable doesn't require air to be a certain temperature; it just needs to feel like that temperature. Even a light breeze will make warm air more bearable. This is why installing a whole-house fan in your home makes sense. A whole-house fan sucks cool outside air in through open windows and pushes hot air into the attic. It forces ventilation through the house, resulting in lower temperatures and greater comfort. The breeze moves over your skin, making you feel cooler yet using only a fraction of the energy an air-conditioner uses. Instead of using air-conditioning to cool the air to 70°F, a whole-house fan makes it feel like it is 70°F, even when the air is above 80°F.

Whole-house fans work best in moderate climates, places where the air might be cool but the sun makes it feel hot. You'll be able to use a whole-house fan for most cooling and save the air-conditioner for only the hottest times, or use the whole-house fan in the morning and early evening hours, when the air is cool, and turn the air-conditioner on for a few hours during the hottest part of the day.

Green $pecs

Overall Rating

Difficulty

Green Benefits

What will this project do for your home?

The sun beats down on your home all day, slowly heating it up. Simply opening a window is not always enough to flush hot air out of the house and replace it with cooler air. A whole-house fan speeds up this air exchange by drawing in a constant supply of cooler, outside air. In just 20 minutes, a home feels cooler, and using the fan means saving energy since you're not using the air-conditioner. For the hot hours of

Whole-House Fan

Installed in the highest ceiling in your home, a whole-house fan draws cooler air from open windows through your home and expels trapped hot air through the attic.

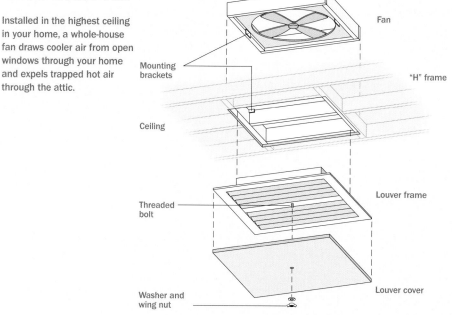

Mounting brackets

Fan

"H" frame

Ceiling

Louver frame

Threaded bolt

Louver cover

Washer and wing nut

the day when you still have to use air-conditioning, a solar attic vent will get rid of trapped hot air (that is sometimes as hot as 150°F) in the attic, likely lowering your air-conditioning bill by 30%.

What will this project do for the Earth?

Air-conditioning consumes 183 billion kWh of electricity in the U.S. every year. Using a whole-house fan could reduce your air-conditioning usage by 30%. If every home installed one, it would save enough energy to power more than 59 million homes. These potential savings would reduce the amount of greenhouse gases released into the air, roughly equivalent to taking 722,000 cars off the road or planting 205 million trees.

Will you need a contractor?

Installing a whole-house fan requires cutting a 30-in. × 30-in. opening in the floor of the attic, between the existing joists. For the avid do-it-yourselfer, it's a fun weekend project. If you've barely held a saw before or if electric wiring scares you, call a handyman, electrician, or contractor to help.

What are the best sources for materials?

Whole-house fans can be purchased online or through building-supply warehouses. Your local mechanical or air-conditioning contractor will also be able to provide and install them.

Typically, a whole-house fan is installed into the highest part of the ceiling below an attic. A central hallway location is best so air is pulled in evenly around the entire house. The top of a stairwell is often the perfect location. To determine the best size for your whole-house fan, take the square footage of your home (not including the basement or garage) and multiply that number by three. A 2,000-sq.-ft. home will need a fan capable of around 5,400 cu. ft. per minute (cfm) to 6,000 cfm at high speed.

Since whole-house fans pump hot air into the attic, you need to make sure you have enough venting around the attic and roof. To calculate the proper amount of venting, divide the size of the fan (in cfm) by 750. For example, a fan that runs 5,000 cfm (5,000 / 750) needs 6.67 sq. ft. of venting.

Remember that most vents are screened, so only about half of the air gets through. You should double the recommended venting numbers just to be safe. Relying on existing attic vents is usually not the best solution. In addition to the whole-house fan, consider installing an automatic attic vent, which helps expel to the outdoors any hot air brought into the attic.

How an Attic Fan Works

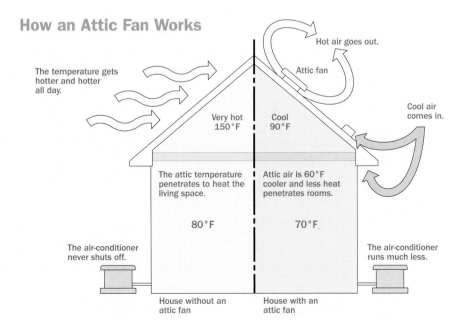

Hot air goes out.

The temperature gets hotter and hotter all day.

Attic fan

Cool air comes in.

Very hot 150°F | Cool 90°F

The attic temperature penetrates to heat the living space. | Attic air is 60°F cooler and less heat penetrates rooms.

80°F | 70°F

The air-conditioner never shuts off.

The air-conditioner runs much less.

House without an attic fan | House with an attic fan

A solar-powered attic vent requires no electricity and automatically exhausts hot air outside. The combination of whole-house fan and solar attic vent does have limitations. The combination of the two can only cool the house to the temperature it is outside and can bring dust and pollen inside. Unlike the air-conditioner, a whole-house fan does not dehumidify the air, so it may not be a good solution in humid places.

A whole-house fan is best used when the outside air temperature is below 82°F. You'll also want to turn off the air-conditioning and operate the whole-house fan during:

- **Spring:** When the sun is warm but air is cool, run the fan to carry the cooler air through the house.
- **Summer:** In the morning, before it gets too hot outside, the fan will precool your home. After a hot day, run the fan again at night to pull cool nighttime air throughout the home. The fan can remain on until morning.
- **Fall:** On days it is warm (but not too warm, so that the air-conditioner is needed).

Most whole-house fans offer two speeds. The high speed is ideal for a quick cooldown. It takes 20 minutes for the fan to replace all of the air in your home. The lower setting is much quieter and used once the home has cooled down enough to be comfortable.

Whole-house fans can be turned on with a simple pull chain, but ask your contractor if a switch can be installed instead. It requires a little more effort but allows you to put the switch in a convenient location. Timer switches are also available, allowing you to turn on the fan for a set period of time.

Some people complain about the noise a whole-house fan makes while running. Select a model with more blades (the more blades, the quieter the fan), and a welded, metal frame, and ask the installer to use foam or rubber spacers to isolate the fan from the structure. This will reduce sound and vibration issues.

While the contractor is in the attic, have him or her caulk around all of the protrusions in the attic floor (electrical boxes, attic ladders, recessed lights, etc.). This helps prevent hot air from sneaking back into your home through little gaps.

How much maintenance will be required after installation?

Installing a whole-house fan creates a large hole in the ceiling of your home. When not in use during the winter months, heat will escape through that hole, so you'll need to place an oversized piece of rigid insulation board over the fan once the weather cools down. Fan covers may also be available for your particular fan, so check with the manufacturer. Put a warning label on the fan switch, reading, "Whole-house fan: Remove cover before operating." This will remind you to take that cover off once the warm weather returns. Otherwise, you risk damaging the fan.

A whole-house fan only cools the home if the windows are open. Close windows in unused areas of the house to increase the air movement in the other rooms. Never operate the whole-house fan with all of the windows closed, as you may pull gas fumes from your heater or hot water heater (this is called a backdraft). Although this might seem obvious, never operate your whole-house fan while a fire is burning in the fireplace or you risk pulling smoke throughout the home.

How long will the project take to accomplish?

Installing a whole-house fan and solar attic vent is different for each home depending on the size of the attic and its accessibility, but the installation is typically done in a single day.

KEVIN SAYS Many of us take the A/C for granted by just setting our thermostats and forgetting about it altogether. The whole-house fan concept is worthwhile because it provides a cool, cheaper fresh-air alternative to A/C.

What is the capital cost? $$

To do this job right, you will need to purchase a whole-house fan and a solar-powered attic fan. The whole-house fan ranges in price from around $200 for a simple 24-in. model with a cover to fancier models with remote-control louvers for roughly $800. The solar-powered attic fan will cost another $350 to $450. If you are not particularly handy you'll need to pay for installation of both fans, which can cost $450 to $1,000. All in all, the entire job should cost around $1,000 to $2,250, which is just a little more than putting in a few good-size window A/C units.

RESOURCES

Jet Fan
www.jetfanusa.com

Solatube SolarStar
www.solatube.com

SunRise Solar
www.sunrisesolar.net

What financial resources are available? $$$

Whole-house fans fall under the American Recovery and Reinvestment Act and are subject to a tax credit for 30% of the purchase price, provided you have not collectively spent more than $1,500 on other improvements. Many utilities (almost all of them in California) offer significant rebates of up to $300 for whole-house and attic fans. Altogether you could receive as much as $500 toward your retrofit.

What is the monthly cost or savings? $$

Federal government surveys show average A/C usage at 16% of an energy bill or $140 per year. Keep in mind these estimates are roughly 10 years old, so $200 to $250 would probably be more accurate, allowing for increases in energy costs. These numbers are, of course, much higher in the South and Southwest. The use of whole-house and attic fans won't eliminate the need for A/C but could reduce costs by as much as 50% if humidity is not a big factor. A savings of $125 annually or about $10 monthly would have your retrofit paid for in about eight years.

What is the long-term home value? $

Bank appraisers don't give value to whole-house fans when appraising properties, and many real estate agents scratch their heads when they see or hear about them for the first time. Documenting the energy savings may intrigue a buyer, provided that the fan has been installed in an aesthetically pleasing manner and it doesn't make a lot of noise.

THE BOTTOM LINE is...

If you live in a dry, moderate climate, a whole-house and attic fan is a great alternative or supplement to air-conditioning. The active interest from the public is motivating manufacturers to make quieter, less visually obtrusive units at lower prices. Installation won't really pay for itself quickly, but you can enjoy fresh air and feel good about saving the environment. ■

Insulate your walls and attic

ERIC SAYS Heating and cooling your home is expensive, consuming 50% to 70% of the total energy used in a home. In the winter, heated air rises and can travel through the home, up to the ceiling and attic, and out into the atmosphere. In the summer, heat flows naturally into cooler spaces, making your home warmer. By insulating your attic (and any new walls you remodel), you can save a large portion of conditioned air and keep it where it belongs.

Green $pecs

Overall Rating

Difficulty

Green Benefits

What will this project do for your home?

Insulating an attic and surrounding walls will have an immediate effect on the comfort inside your home. A well-insulated home uses 30% to 50% less energy than a home without the minimum amount of insulation. Since your heater and air conditioner will not have to work as hard to keep the home comfy, it also indirectly reduces the need for maintenance and repairs.

What will this project do for the Earth?

Most older homes in this country are poorly insulated. There are more than 50 million underinsulated homes in the U.S., wasting the equivalent of about 2 million barrels of oil every day in lost energy. That is more oil than we import from Saudi Arabia each day.

Will you need a contractor?

No matter what type of insulation used, adding insulation to new walls or an existing attic is relatively simple. It requires access to your attic and a bit of crawling on your hands and knees. But if the thought of crawling around a hot, dusty attic doesn't appeal to you, a licensed and bonded contractor could do the job easily.

Insulation Zones

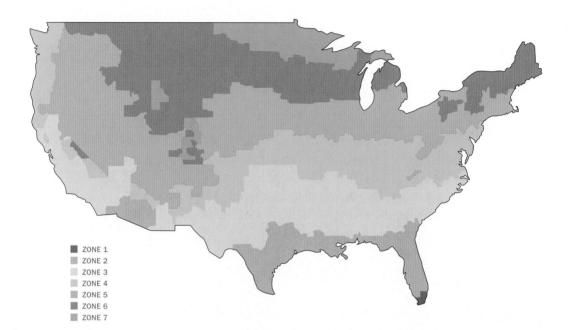

- ZONE 1
- ZONE 2
- ZONE 3
- ZONE 4
- ZONE 5
- ZONE 6
- ZONE 7

ZONE	WALLS	ATTIC	CEILING	FLOOR
1* Miami Hawaii	R13–R15	R30–R49	R22–R38	R13
2 Houston New Orleans	R13–R15	R30–R60	R22–R38	R19
3 Atlanta Little Rock Los Angeles	R13–R15	R38–R60	R22–R38	R25
4 Raleigh Nashville Portland	R13–R15	R38–R60	R30–R38	R25–R30
5 Pittsburgh Reno	R19	R38–R60	R30–R38	R25–R30
6 Fargo Casper	R21	R49–R60	R30–R60	R25–R30
7** Jackson Hole Alaska	R21	R49–R60	R30–R60	R25–R30

*Zone 1 includes Guam, Puerto Rico, and the Virgin Islands.

**All of Alaska is included in Zone 7 except for the boroughs of Bethel, Dellingham, Fairbanks, North Star, Nome, North Slope, Northwest Arctic, Southeast Fairbanks, Wade Hampton, and Yukon-Koyukuk. Add five to the suggested R-values for these areas.

What are the best sources for materials?

Insulation can be purchased from any home center store or large hardware retailer. Measure the areas to be insulated before you go to the store so you can purchase the correct amount of material. If using a contractor, he or she will likely be able to purchase the insulation for you at a better cost.

Insulation is the reason you wear a coat in the winter; the coat holds in your body heat and prevents the cold from reaching you. It works by holding in warm or cool air, and the ability of a material to insulate is measured by what is called R-value. The higher the R-value a material has, the greater the insulating ability. Technically speaking, every material has at least some R-value, even if it is very low (like glass).

Depending on where you live, the amount of R-value the insulation in your home should have varies (see the map on the facing page). The Department of Energy website also has a helpful Insulation Calculator (www.ornl.gov/~roofs/zip/ziphome.html). And remember, insulation is like chocolate: The more you have, the better. Install as much insulation as your walls and attic can hold and it will make a drastic difference in your comfort. A typical wall (made of 2×6 wood studs) is thick enough to hold R-19 fiberglass batts. You also probably need to insulate (or add insulation) if your home was built before 1981 (when standard energy codes went into effect); if you have rooms that are stuffy in summer or freezing in winter; or if you are currently remodeling—it's the best opportunity to add insulation easily.

Types of insulation There are many types of insulation, and each has a different R-value (see p. 129). Most people are familiar with the "pink stuff," but insulation comes in four major types:

- **Batts:** Batts are fluffy blankets of insulation and the most common type available. They come in rolls and install easily by filling the space between the studs in a wall or the joists in an attic or ceiling. They are available in pink fiberglass, rock wool, or cotton. If you choose fiberglass, be sure it is formaldehyde–free. As an option, batt insulation comes "faced" with kraft paper or foil. Facing is an outer layer that adds a little extra R-value to the insulation. Use faced batts in areas where you need the extra R-value.

- **Loose-fill insulation:** As the name implies, loose fill consists of small, shredded pieces of insulation that are blown into place using a special blower. The small bits fill up cavities, spaces, and hard-to-reach areas like attics where installing other types of insulation is difficult. If installed into walls, loose fill has to be held in place by netting (to keep it from falling out), or be slightly sticky to stay in place while waiting for the wall to be finished.

- **Foam-in-place insulation:** Foam-in-place insulation is similar to loose fill in that special equipment is needed to install it. It expands like shaving cream and fills every nook and cranny, creating a super-insulated area. It's perfect for walls, especially around windows and doors.

  A wide spectrum of foam-in products is available, including some toxic chemical concoctions. The greenest and healthiest choice is soy-based foam. Spray-in-foam insulation is my top pick since it fills the entire wall cavity, although price may be prohibitive. Be sure to use a foam product that uses water as the blowing agent, not chemicals.
- **Rigid-foam insulation:** Rigid-foam insulation is a stiff board, not fluff or foam. Although it is usually more expensive, it has up to double the R-value per inch of batt insulation and is used in places where space is limited. Rigid boards are often used in remodels where it is difficult to access certain areas or there is not enough space for batts. Look for manufacturers of recycled content boards, or polyisocyanurate (polyiso for short) boards.

Insulating the attic Check your attic and measure the depth of the insulation. If you have less than 11 in. of fiberglass or rock wool or less than 8 in. of cellulose, then add more. You want between R-30 and R-60 in your attic. The key to insulating the attic is to install it on the

Loose fill is available in fiberglass, rock wool, or cellulose. Look for formaldehyde–free products, or natural products, such as cellulose from recycled newspapers.

floor of the attic, not the roof. You want the insulation to prevent heat and air-conditioning from going up into the attic entirely.

In many homes, loose-fill insulation is the easier type to install. You'll want to make sure the insulation is installed evenly around the entire attic. Avoid blocking any of the soffit vents around the edges of your roof. Do not allow the insulation to come in direct contact with recessed lights or any mechanical equipment. Special covers are available for recessed lights.

If batts are used in the attic, place the first layer between the floor joists. They should fit snugly. Place the second layer of insulation perpendicular to the first. This covers the top of the joists and ensures there are no gaps.

Don't forget to cut a piece of rigid board for the attic access panel. Cut it to fit and glue it on the backside of the access door. If you have an attic ladder, keep a large, loose board in the attic that you can slide over the opening before you close up the ladder.

Insulating new or remodeled walls If you're remodeling or building new walls, adding insulation is easy and is only needed on the exterior walls of your home, not between rooms. While the walls are open and accessible, this might be the only opportunity you have to add insulation easily.

Insulating existing walls It is incredibly difficult to insulate an existing closed wall. Holes near the ceiling must be cut into the wall every 16 in. to allow access to the air space between each stud. Loose-fill insulation is then pumped into the wall. The holes are patched when complete. It is a messy and ineffective way to insulate, but it is less wasteful than ripping off drywall and replacing it later.

My advice is to wait. If you plan to remodel or replace your home's exterior siding in the future, you'll have access to the wall and that might be a better time to insulate. If you are unsure if a wall is insulated, take the cover off an electrical outlet and peek in with a flashlight.

Insulation tips and techniques I have seen countless installations where insulation was installed incorrectly. Never pinch or compress insulation. The fluffiness and air is what helps the insulation

The greenest and healthiest choice of batts is recycled cotton, such as formaldehyde-free recycled denim insulation.

work well. Packing the insulation down like clothes in a suitcase will prevent it from working. Avoid gaps or voids.

Insulation should completely fill the space between joists, fit snugly at the sides and the ends, and be cut to fit around plumbing, wiring, and vents. Make sure insulation continues to the edges of the studs. For large pipes, peel batt insulation in half and place one layer behind the pipe, the other in front. When insulating around recessed light fixtures, remember that you can only cover the light with insulation if it is marked "IC," meaning it is safe for insulation contact. Otherwise, use a cover or baffle to keep the insulation from coming into direct contact with the hot light.

Always wear gloves, eye goggles, a breathing mask, and a long-sleeved shirt when installing or coming into contact with insulation. Never touch your face or rub your eyes while handling or being around insulation. Bare skin and insulation do not get along.

If you are planning on reroofing, consider installing a radiant barrier below your roof. Its reflective surface will cool your attic even more, working with your attic insulation. Attics above a garage don't need to be insulated since they are not usually heated or cooled. Unless you spend time in your garage or need it cool, skip the insulation.

How much maintenance will be required after installation?

Once the walls are closed up, you will not have a chance to inspect the insulation. By installing it right the first time, you can save yourself future headaches.

Since the attic remains open and accessible, you should check annually for roof leaks, mold, or low spots in the insulation. Birds, rodents, and even the wind can disturb insulation, and you want to make sure that it has not shifted into direct contact with any recessed lighting.

How long will the project take to accomplish?

Insulating an attic takes a few hours and can be done anytime (although you might not want to do it on a hot, summer day!). Insulating walls can only be done once the plumbing, wiring, and ducts are fully installed. Insulating one wall can be completed in less than an hour.

R-Values of Insulation Types

Different insulation materials have different R-values. The standard insulation, fiberglass batt (the pink stuff), has around 3.14 of R-value for every inch. But you can see from the chart below that polyisocyanurate has more than double that R-value per inch.

INSULATION MATERIALS	R-VALUE PER INCH OF MATERIAL
Fiberglass batt	3.14
Fiberglass blown (attic)	2.20
Fiberglass blown (wall)	3.20
Rock wool batt	3.14
Rock wool blown (attic)	3.10
Rock wool blown (wall)	3.03
Cellulose blown (attic)	3.13
Cellulose blown (wall)	3.70
Vermiculite	2.13
Rigid fiberglass	4.00
Polystyrene	5.00
Polyurethane	6.25
Polyisocyanurate	7.20

SPACE REQUIRED FOR R-VALUES USING STANDARD FIBERGLASS BATTS	
R-value	Thickness needed
R-13	4.1 in.
R-19	5.9 in.
R-22	6.9 in.
R-24	7.5 in.
R-25	7.8 in.
R-30	9.3 in.
R-38	11.7 in.
R-42	12.9 in.
R-49	15.0 in.
R-60	18.3 in.

KEVIN SAYS Most homes built after 1970 have some form of insulation, and many older homes have already been retrofitted—at least in the attic. If your home is one of the homes left out, then insulation is certainly worth considering.

It's possible that your older insulation can be upgraded to a higher R-value to help save even more energy. Plus, insulation keeps the home temperature a little more regulated and can add noise dampening as well.

What is the capital cost? $$$

The cost of insulation will vary depending upon location and material, but figure anywhere from $0.50 for handy DIYers to $2.25 per sq. ft. for fully blown or installed. That means an attic with 2,000 sq. ft. of space will cost about $1,000 plus a bit of your time, or you can sit back in front of the TV and get the nontoxic spray-in foam done for about $4,500.

What financial resources are available? $$$

Insulation material costs are reimbursable up to 30% with a $1,500 limit from the federal government. Many local utilities offer rebates and credits. Georgia offers up to $1,900 for contractor installation, and California's Pacific Gas and Electric will pay $150 per 1,000 sq. ft. Typical recoup might be as much as $400 to $650 on average. The government mortgage organizations Fannie Mae and Freddie Mac are now allowing certain energy retrofits such as insulation to be financed on top of appraised value. This won't save you money but may make it easier to afford.

What is the monthly cost or savings? $

Savings vary based upon energy usage and the quality of insulation you install, but the Department of Energy figures a savings from 4% to as much as 22% on your energy bill. Obviously the greater savings go hand in hand with the more expensive insulation, so on average you can recoup the cost of insulating in 8 to 10 years.

What is the long-term home value? $$

Few bank appraisers spend extra time digging in the attic or basement looking for insulation. And since they can't check comparable sales for the addition of insulation, they generally don't use it as a line item for value. Real estate agents will likely point out that the house is quieter and well insulated as a feature, but it's difficult to justify a higher asking price for a higher R-value when compared with another insulated home. But an older home that has been insulated may recoup the cost in sales price versus an uninsulated home that is also being considered by a buyer.

THE BOTTOM LINE is...

Comfort is reason enough to insulate a house. Reducing noise and keeping the temperature steady makes for a more comfortable home, and you can feel good that the heating and cooling you pay for isn't going to waste. If you are going to be in your home for a while, you'll probably make back your investment. Now is a good time to insulate since the government and local utilities will help with the cost. ■

23

Upgrade your appliances

ERIC SAYS There are really two price tags when buying appliances: what you pay to take it home, and what you pay for the energy and water it uses. Upgrading to Energy Star appliances (look for the black and yellow label) helps you figure out how to manage that second cost.

Energy Star, a program developed by the U.S. Environmental Protection Agency and Department of Energy, is a certification for the energy efficiency of appliances. An Energy Star–rated appliance meets a minimum standard of energy performance. The four largest energy-consuming appliances in your home—the refrigerator, washing machine, clothes dryer, and dishwasher—are the best places to upgrade to new Energy Star models for impressive energy savings. The Energy Star website (www.energystar.gov) maintains a comprehensive list of rated appliances, which is a good place to start researching upgrades.

Green $pecs

Overall Rating

Difficulty

Green Benefits

What will this project do for your home?

On average, any standard appliance you upgrade to an Energy Star model will reduce its energy use by 30%. For example, the refrigerator is the largest single energy user in your home. By replacing a 1990 or older model with a new Energy Star model, you'll save enough electricity to light your home for four months. The average family does around 400 loads of laundry a year. Advances in technology since 1999 ensure significant energy savings if you upgrade that washer and dryer. Energy Star–rated dishwashers use up to half the water and more than 40% less energy than conventional models.

What will this project do for the Earth?

More than 47 million outdated refrigerators are still in use in the United States. If all these were upgraded to Energy Star units, it would save enough energy to power 14 million homes every year.

Will you need a contractor?

You probably won't need one for installing a new refrigerator unless it has an icemaker or water dispenser, in which case you'll need a handyman or plumber to connect it to a water line. Replacing the washer and dryer only requires a few screw-in connections, such as the hot and cold water hoses and dryer vent hose. The dishwasher will require some help. One mistake in the installation could lead to future leaks.

What are the best sources for materials?

Home centers, department stores, and manufacturers online are the places to start. Be aware that just because a manufacturer has one Energy Star–rated appliance model doesn't mean all of their models are Energy Star rated. Check each model before buying, and remember that the larger the appliance, the more energy it uses. Select the smallest appliance that meets your needs, but keep in mind that one large appliance is more efficient than two smaller models. If buying a new, larger fridge allows you to unplug that old backup freezer, you'll save more energy. European-brand appliances are often more water and energy efficient, and some companies manufacture their products in the U.S.

When to Upgrade that Appliance

Refrigerator: Recent national standards have reduced the energy use of refrigerators to less than one-third of 1973 models. Current Energy Star refrigerators use half the energy of models made before 1993. And just since 2001 the energy standards have dropped by 40%. Seriously think about replacing any refrigerator manufactured before 2001 with a new Energy Star–rated one.

Washing machine: Replace any washing machine manufactured before 1999 with an Energy Star–rated model. Since a new one saves both energy and water, replacing an inefficient washing machine provides the biggest savings over any other appliance.

Clothes dryer: In terms of energy savings, it is not worth replacing a clothes dryer until it reaches the end of its useful life. Rather than upgrading it, just use a lower heat setting and clean the lint filter after every use. Better yet, use a clothesline (see p. 80).

Dishwasher: Replace any dishwasher manufactured before 1994. Current Energy Star dishwashers use more than 40% less energy than older, inefficient models.

Choosing and operating a refrigerator
Refrigerators with an upper or lower freezer are (generally) more energy efficient than side-by-side models. An automatic defrost refrigerator with side-by-side freezer and door dispenser uses nearly 40% more energy than a model with a top freezer and manual defrost. Automatic icemakers and door dispensers use slightly more energy, too.

Avoid placing a new refrigerator near a stove, dishwasher, or heating vent. Set your refrigerator between 35°F and 38°F and your freezer at 0°F.

Choosing and operating a washing machine Washing machines with a horizontal axis (typically front loading) are more efficient than top-loading washers because front-loading washers don't need to fill the tub completely, using much less water. Select a washing machine with options for energy-saving wash cycles (and use them!). A hot-water wash with a warm rinse uses 5 to 10 times more energy than using cold water alone. From 70% to 90% of the energy used in a washing machine goes toward water heating. Less hot water equals less energy used.

If you choose a standard top-loading machine, look for a model with adjustable water level options. Adjust the water level to match the size of the load. Full loads save more energy and water than doing several small loads—a washing machine uses the same amount of energy regardless of load size.

Several manufacturers offer all-in-one washer and dryers. These single units take up less space than side-by-side units and use much less water and energy to operate.

Choosing and operating a clothes dryer Energy Star does not label clothes dryers. The best way to save energy on drying is to use a low heat setting and clean the lint filter after every load. Selecting a model with a moisture sensor saves energy by automatically shutting off when clothes are dry. Not only does this save energy, but it also reduces the damage to your clothing from overdrying.

Choosing and operating a dishwasher Look for Energy Star dishwashers with booster heaters and air drying. Only operate the dishwasher when full. Use the energy-saving options, such as short wash, and skip the prerinse.

Don't ignore those black and yellow stickers! An Energy Star label shows the amount of energy a particular model uses compared with other similar models. The bottom of the label estimates how much it will cost to run the appliance.

How much maintenance will be required after installation?

A new appliance should last 12 to 15 years with minimal maintenance. Use a vacuum to clean refrigerator rear coils every three months to improve efficiency. Check the gaskets around the door for leaks. Manually defrost the freezer if you have more than 1/4 in. of ice buildup. Regularly inspect seals around the washing machine and dishwasher for leaks.

Keep all of your appliance manuals, warranty cards, and paperwork together in case you ever need them. I find the best place for those things is in a large plastic bag taped to the back of the appliance.

How long will the project take to accomplish?

Installation of an appliance can be done in a couple of hours if you are replacing one of the same size. Donate old appliances to your local salvage yard, put it on craigslist (www.craigslist.org), or drop it off at Goodwill (www.goodwill.org). Energy Star also provides a recycling program for refrigerators (www.energystar.gov/recycle).

KEVIN SAYS I have bought my fair share of appliances and remodeled several kitchens over the years. While I appreciate the fact that energy efficiency has improved greatly over the years, it has rarely been my first priority in selecting appliances. Most people aren't looking to replace perfectly good appliances until they wear out, but new energy-efficient appliances can show us what we are missing. The technology is impressive. My new dishwasher uses the heat from the hot water to dry the dishes. It takes twice as long, but I feel good about reusing the energy. And if there's a chance to get shiny new toys, save the planet, and save money as well, it's certainly worth a look.

What is the capital cost? $$$

The cost of appliances varies greatly depending upon brand, capacity, and features, but nice units can be bought for reasonable prices and financed through most major retailers. Refrigerators start at around $750 and can cost more than $2,500 for super-fancy models with lots of features that diminish the energy savings (like those with TVs in the door). Washers can run from small, compact units for $500 up to $1,500 for large-capacity front-load models. Dishwashers cost as little as $285 but can go as high as $1,500. Noise is often a factor in cheaper models.

Paying more won't necessarily assure you of better energy savings, although specific energy-focused units can raise the price. If you are willing to pay more for a super energy-efficient fridge, you can expect to spend as much as an additional $500 to $700 for the privilege of being more energy efficient.

What financial resources are available? $$

Select Energy Star appliances carry rebates ranging from $50 to $100. Also, big-box stores advertise specials just about every week. Serious bargain hunters might look into discount outlets for major retailers such as Sears, where you can pick up brand-new appliances with a small scratch or dent for 25% to 35% off sticker price and still get the rebate.

What is the monthly cost or savings? $

Most Energy Star estimates show roughly 30% energy savings when comparing modern, energy-efficient appliances with those more than 10 years old. With standard low-feature units, a typical homeowner buying a new washer, fridge, and dishwasher can save as much as $150 to $200 per year depending on usage. Obviously the savings alone are not enough to cover the cost of the appliances in less than about 15 years, but at least you get the convenience of all-new appliances.

There is a point of diminishing returns in paying more for lower energy-rated brands. Even though a Sun Frost refrigerator is great on energy savings, its Energy Star usage is rated at roughly $17 annually versus a typical top-freezer name-brand model of the same capacity that is rated at $44 per year. With the $27 annual energy savings, it will take you nearly 20 years to recoup the difference in cost.

What is the long-term home value? $$

Appliances have little impact on home value unless they are old and need replacing, or are built-in and upper-end brands. If investing in top-of-the-line appliances versus basic models, you may be able to get a higher price provided that the kitchen is put together in an elegant, usable manner. Bank appraisers will give some additional value to a fully upgraded kitchen.

THE BOTTOM LINE is...

If you are sitting on appliances that are seven to eight years old or more, upgrading is worth considering. Even if you don't replace everything, you can help the planet and your wallet by buying something a little sooner than needed since it won't be long until they start breaking down. If you are remodeling, it is the perfect time to go all new. Who wants a brand-new kitchen with old, beat-up appliances? Just make sure the old ones are disposed of responsibly. ■

RESOURCES

Energy Star refrigerator recycling program
www.energystar.gov/recycle

Refrigerators:

Absocold
www.absocold.com

ConServ Equator®
www.equatorappliance.com

MicroFridge®
www.microfridge.com

Summit Compact Refrigerators
www.summitappliance.com

Sun Frost
www.sunfrost.com

Dishwashers:

Ariston
www.aristonappliances.us

Asko
www.askousa.com

Bosch
www.boschappliances.com

Danby®
www.danby.com

Miele
www.mieleusa.com

Upgrade your fireplace

Green $pecs

Overall Rating

Difficulty

Green Benefits

ERIC SAYS A fireplace is the symbol of an inviting and luxurious home, but all that symbolism can come at a steep price. Having a fireplace actually adds to the cost of heating a home. When not in use, a fireplace allows heat to fly up the chimney and cold air to leak in. Even when it is in use, it draws in and burns up heated air, and then cold air is pulled in through cracks in the walls to replace it, and most of the heat it produces rises up through the chimney. A fireplace really only heats the area around it, and having both a fireplace and mechanical heating system compete to keep you warm only reduces their energy efficiency. The average efficiency of a fireplace ranges from −20% (yes, that's right—negative, meaning it loses more heat than it produces) up to only around +20%.

If you want to keep your existing fireplace, then improve its efficiency by upgrading it with grates, heat exchangers, and glass doors, or look into sustainable fuels. You'll see a dramatic improvement in your heating bills and comfort.

What will this project do for your home?

A traditional masonry fireplace burns up to 300 cu. ft. per minute of heated room air, making a home cooler, not warmer, and adding 8% to 10% to a heating bill. Leaving a fireplace damper open when a fireplace is not in use increases heating and cooling energy use by 30%. Most of that wasted energy could be saved by upgrading a fireplace, and switching to an alternative fuel such as alcohol or sugar will prevent toxic wood smoke from entering your home, making for a healthier environment indoors and out.

What will this project do for the Earth?

Fireplaces are the largest intentional source of heat loss in the home and cost Americans more than $6 billion a year in wasted energy costs. If all of the 33 million fireplaces in the U.S. were sealed properly, it would save enough energy to heat 5% to 10% of homes in the U.S. each year. Wood-burning fireplaces are also a major source of air pollution and their smoke is carcinogenic. On a cold winter evening, wood smoke can contribute up to 80% of the particulate emissions in your neighborhood.

Will you need a contractor?

Making upgrades to an existing fireplace is tricky. Since you are literally playing with fire, it's best handled by professionals. Your local fireplace company can perform the upgrades.

What are the best sources for materials?

Home center retailers, hardware stores, and fireplace accessory showrooms stock inserts, grates, and doors. You'll find better deals and more information online. Internet resources like www.fireplacemall.com are great for one-stop shopping on grates, firebacks, and dampers.

Fireplace accessories Installing glass doors prevents room air from blowing up the chimney when the fire is low or the fireplace is not in use. Another useful accessory is a fireplace heat exchanger, also known as a blower door, which sits below the fireplace doors and circulates the room air through the fire and back into the room. Look for a unit made of durable materials to resist the corrosive effects of the flames.

When choosing a grate for the fireplace, look for one that cradles the wood and allows air to reach the underside of the wood, which allows heat to radiate into the room instead of just the chimney. Similarly, a fireback, a cast-iron or stainless-steel panel that sits against the back wall of the fireplace, reflects the heat of the

Fireplace Heat Exchanger

A fireplace heat exchanger draws cool air in from the room and blows out air heated by the flames, greatly improving the energy efficiency of a fireplace.

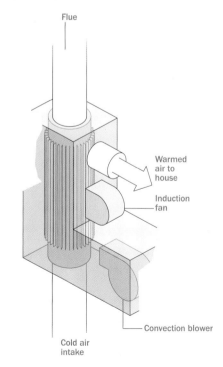

Flue

Warmed air to house

Induction fan

Convection blower

Cold air intake

EPA-Certified Fireplaces

The U.S. Environmental Protection Agency (EPA) certifies low-polluting fireplaces that produce a low amount of smoke and a minimal amount of ash. Similar to the Energy Star rating (see p. 131), the EPA certification is a trusted seal, and has certified a wide range of fireplace types (www.epa.gov).

You can also upgrade an existing fireplace with an EPA-certified insert. A fireplace insert is an energy-efficient unit that fits into the existing fireplace opening and requires a metal flue liner for the chimney.

flames back into the room and protects the masonry, making a notable difference in efficiency.

A damper is a metal door located at the bottom of the chimney that is supposed to keep air from coming in or out through the chimney. Dampers are typically made of a thin metal plate and are notoriously leaky and usually do a poor job of insulating. The solution is to install a top-sealing damper at the top of your chimney, creating a weather-tight seal.

Try a new fuel You may want to skip the gas or wood entirely and try an eco-friendly, alternative-fuel fireplace such as one that burns ethanol. Liquid ethanol, a renewable fuel most often made from sugar, potatoes, or corn, is poured into a metal firebox and lit like a camping lantern. The best part about this type of fireplace is that it can be placed anywhere in the home—no chimney is needed since the fuel burns away completely. And no chimney means no need to comply with any building codes.

Pellet stoves are highly efficient, provide a great deal of heat, and burn compressed wood pellets instead of logs. The pellets resemble rabbit food and do not have a visible flame.

A wide variety of gel- and alcohol-burning fireplaces can be found on the Internet. Many are freestanding or can be incorporated into existing fireplaces.

A surprisingly green alternative to wood is manufactured fire logs. They are typically made from sawdust, vegetable wax, and other recycled materials, such as nutshells, cardboard, and even discarded coffee grounds. Fire logs used to contain petroleum but they now produce less than half the amount of carbon dioxide and 40% less smog.

How much maintenance will be required after installation?

If you use your fireplace even just a few times a year, it should be professionally inspected and cleaned annually. Over time, smoke lines the chimney with creosote, which poses a fire hazard if not removed. The Chimney Safety Institute of America (www.csia.org) provides a list of certified chimney sweeps around the country.

Several eco-friendly fireplaces burn ethanol instead of wood or gas. The fuel burns away completely, eliminating the need for a chimney. This type of fireplace often comes on wheels and can be moved around the house.

How long will the project take to accomplish?

The installation of glass doors can be done in an afternoon. Installing a new insert could take a few days depending on how complex the location is.

KEVIN SAYS I am an everyday fire person when the weather gets a little chilly, and I've switched over to manufactured logs partly out of convenience. Keeping wood dry and ready and lugging it in the house every night were enough of a chore that I now just pick up a box of logs at the market once a week. There are now plenty of affordable fireplace options that mean you can get the most out of the one you own with minimal environmental impact.

What is the capital cost? $$

An insert with glass doors that will efficiently transfer heat and prevent energy from leaving the room costs around $1,800. Bringing in a freestanding smokeless fireplace costs at least $350 and can go well beyond $3,000 for high-design models. Grates, firebacks, and dampers run from $75 to $250 each. Inserts and dampers may have labor costs that could run anywhere from $150 to $500.

What financial resources are available? $

The most that the government will cover is 30% of the cost of the materials, up to $1,500 total. A few municipalities offer funds to switch to gas from wood, and I found that the manufacturer Avista, in Spokane, Washington, gives out $100 toward a damper installation if you use a contractor.

What is the monthly cost or savings? $

Installing a damper or glass doors saves anywhere from $10 to $25 per month, depending on where you live. Otherwise, you are essentially swapping fuels. If you are like me—and have a fire every night during fall and winter—you will spend roughly $120 to $160 per month for fuel, be it gel, wood, or logs. If you use an exchanger and are going to have the fire anyway, you could see maybe another $10 to $20 in monthly savings. But replacing a furnace entirely with your fireplace will not likely put you financially ahead of the game.

What is the long-term home value? $

Having a permanent fireplace in the home is always attractive to buyers. The National Association of Realtors states that a fireplace can add as much as $12,000 in resale value, depending on quality and condition. If the fireplace is kept functional and nice looking, you should be able to recoup any costs you spend in making it healthy and efficient. Freestanding units, however, will be assessed as furniture and have zero impact on home value.

THE BOTTOM LINE IS...

It's easy and cost-effective when it comes to making a fireplace eco-friendly. It's not a big expense and the green options don't really require any sacrifice. The new smokeless options even let closed-in apartments enjoy an open flame. It brings me back to the beginning of time when our ancestors said it best: Mmmm—Fire—Gooood! ■

Heat only the water you use

ERIC SAYS Imagine how wonderful it would be never to run out of hot water. Everyone, at some point, has experienced the shock and upset of running out of hot water in the middle of a shower. After all, nearly 20% of all of the energy consumed in the home is used to heat water. Most homes store hot water in a giant tank called a hot water heater. It slowly heats the vat of water until it reaches a set temperature. The water starts to cool down immediately, so the heater runs all day to keep the water hot. If you use too much hot water before the heater has a chance to do its job, you get zapped with a cold shower.

To prevent this from ever happening, many people turn up the thermostat on the hot water tank or buy a larger tank. Doing this wastes an incredible amount of energy by heating more water to a higher temperature than needed. A better solution is investing in a tankless water heater—also referred to as an "on-demand" heater, which has no storage tank and doesn't continually heat a large supply of water all day long. Instead, tankless units instantly heat up only the water being used, providing a limitless supply of hot water. And when the shower shuts off, the tankless heater shuts off, too.

What will this project do for your home?

A conventional hot water heater stores water at a high temperature (usually between 120°F and 140°F) to ensure the water has an adequate amount of heat by the time it reaches you. To prevent burning or scalding, most people run the hot and cold water at the same time to bring the temperature down to a comfortable level. So you are paying to heat the water, only to cool it down again. Since tankless heated water doesn't have to sit in storage, it can be set to a lower temperature. And by installing a tankless heater closer to the bathroom, much of the

Green $pecs

Overall Rating

Difficulty

Green Benefits

How a Tankless Water Heater Works

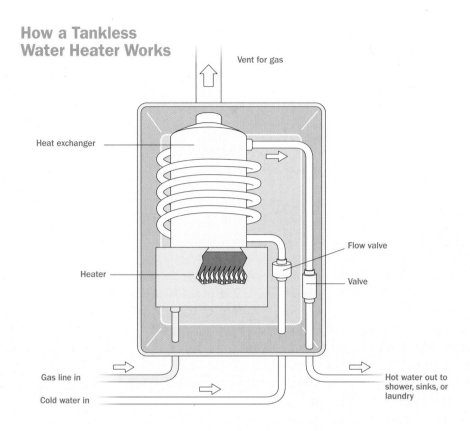

Vent for gas

Heat exchanger

Heater

Flow valve

Valve

Gas line in

Cold water in

Hot water out to shower, sinks, or laundry

water (upwards of 10,000 gal. every year) that is wasted while people wait for hot water to arrive at the tap could be saved.

Tankless water heaters are available in gas and less-efficient electric models. A gas-powered on-demand heater consumes 20% to 50% less energy for water heating than a conventional model. An electric model saves 10% to 20% in energy use compared with a conventional hot water heater.

What will this project do for the Earth?

A conventional tank water heater is second only to heating and cooling in terms of energy usage in the home. In the U.S. alone, water heating consumes 122 billion kilowatt hours of energy a year. By switching every home to a tankless unit, we'd save enough energy to fully power more than 2.2 million homes for a year.

Will you need a contractor?

Installation of any type of water heater involves knowledge of plumbing, gas, and/or electrical lines. This is not something most people could (or should) handle on their own. Depending on where you live, you may need a permit to install a water heater and there could be additional requirements that impact installation. A licensed plumbing contractor should be familiar with the local code requirements. A few years ago, you might have run into issues with local building codes. Today, on-demand heaters are accepted everywhere.

Installing an on-demand heater can cost twice as much as a tank heater, due to the need for a larger gas line and vent. The existing gas line in your home may not be the correct size. Check the manufacturer's requirements and confirm the size of your current gas line before starting any work.

What are the best sources for materials?

The installation of a tankless heater is nearly identical to a conventional tank. You could simply install the tankless unit in the same location as the conventional tank, but ideally it would be installed as close as possible to the main bathroom. The small size of a tankless unit means it can be installed inside a closet, but a tankless heater does require either gas or electricity hookup and a vent to the outdoors.

Electric on-demand units use twice as much energy, but gas units are more difficult and expensive to install. In my experience, an electric unit makes sense if you have solar panels on your roof, have no nearby gas line, or want hot water for a little-used area of the home, such as a guest room or pool house. You can find a variety of models available at your local hardware store, plumbing-supply warehouse, or home center.

The size of an on-demand heater depends on the number of faucets that may need to run at the same time. If the shower sprays 3 gal. a minute and you select an on-demand heater with a flow rate of only 5 gal. a minute, then there will probably not be enough hot water for two showers to run at the same time. You can be careful to run only one at a time, install two (or more) on-demand heaters together to meet your hot water needs, or look into combining a mini tank with an on-demand heater.

Extensively used in Europe for decades, on-demand heaters are now widely available in the U.S. Installing a tankless heater as close as possible to the bathroom reduces the wait for hot water to only a few seconds and can save dozens of gallons of wasted water per use.

Tankless heaters turn on automatically anytime a hot water faucet is opened. When running, a tankless unit makes a lot of noise—some compare it to the noise made by a small jet engine! Because of this, you may not want to install it near a bedroom.

If you are installing a radiant heating system (now or in the near future), then an on-demand heater may not be compatible. A radiant heating system requires a conventional tank water heater for storage and will not tie into an on-demand unit (see p. 257). If you are planning to install solar water heating on your roof, the on-demand heater is not a suitable option because as the water on the roof is heated by the sun, it needs to return back into a conventional hot water tank for storage (see p. 147).

Although you might hear people refer to an on-demand heater as an "instantaneous" heater, it does take a few seconds for water to heat. Plus, the water sitting in the pipes in the wall will be cold, and the farther away the water heater is from the shower, the more cold water has to come out before the hot water arrives. If you are replacing your old tank heater with an on-demand unit, this distance will not change, and you will still wait for the hot water. And regardless of whether you have a tank or a tankless water heater, the amount of energy required to heat up the water is the same. This is simple physics. The on-demand heater will still use energy to heat up your water. The real energy savings in the on-demand heater comes from heating water only once, as you need it. It is not heating a tank of water, letting it cool, and heating it up again.

How much maintenance will be required after installation?

Tankless water heaters typically last for 20 years (compared with 10 years for a standard tank heater). Like any appliance, they have parts that need maintenance and possible replacement. Unlike tank heaters, they rarely leak and cannot flood your home, and a tankless heater is more resistant to mineral buildup in pipes due to the fact that water is always pushed through at full pressure.

How long will the project take to accomplish?

Installation time varies depending on the need for a new gas line, the location of the outside vent, and the connection to the other pipes. Request an estimate in writing before starting any work and verify if a permit will be required and who is responsible for getting that permit.

KEVIN SAYS I don't really think about my hot water heater. It seems durable. A client of mine bought an 80-year-old house with the original coil heater still working. Gaining time by not having to wait for water to heat up would be beneficial, especially if I can help the planet at the same time. However, the cost versus benefit of putting in new tankless technology is questionable.

What is the capital cost? $$

A new tank heater can be bought and installed for roughly $600 to $800, depending upon size and brand. Tankless heaters have come down a bit in price, with the cheapest starting around $700 and topping out around $1,500. Installation costs can be trickier to estimate because it depends on location and whether the heater is gas or electric. Moving pipes or upgrading an electrical panel will be an additional cost. Figure on spending an additional $300 to $700 on top of the cost of the heater, putting your total conservatively around $1,000 to $2,200.

What financial resources are available? $$

There may be tax credits available for switching to a tankless system. A federal law was approved in 2007 that was extended to 2009 for storage water heaters with a federal energy factor (EF) rating of 2.0. This is the highest rating, taking into account recovery, standby, and cycling efficiency. The law allows for up to $300 in tax credits, but there is no guarantee of extension. Small rebates are available through local utilities, but the requirements and size of the rebates vary greatly. None of the rebates or tax credits cover more than about one-third of the cost of a tankless heater.

What is the monthly cost or savings? $

Estimates vary, but replacing a 15-year-old heater with a new tank heater might save as much as 10% to 15% of the energy used, which could translate into roughly $25 to $40 savings per year. At that rate, it would take nearly 20 years to recoup the cost. A tankless heater won't do much better. Tests show roughly 20% to 30% energy savings, but that only translates to about $50 to $80 per year, which means waiting more than 10 years to recoup costs.

What is the long-term home value? $

House buyers often purchase a home warranty to cover appliances such as water heaters. For this reason, most people don't really care how old the water heater is as long as it works and they aren't willing to pay more for a recent upgrade. After checking with appraisers in various parts of the country, we determined that there is no comparable data to support additional home value for a tankless heater.

THE BOTTOM LINE is...

If you want the convenience of instant hot water, then you might consider a tankless water heater, but the cost doesn't justify replacement just to save the planet. If you recently changed your water heater, it's best to leave well enough alone. But if your water heater is at least 10 years old, you should probably change it for safety and energy reasons. For now, let cost be your guide, but keep in mind that new technologies are just around the corner. ■

Install a solar hot water heater

ERIC SAYS If you've ever had to go up onto your roof, you've probably noticed how hot it gets up there. As the sun beats down on the dark shingles, a roof can reach temperatures of up to 130°F. Imagine capturing that free heating for use in your home. That is the basic idea behind solar hot water heating.

After heating and cooling, water heating is the second-largest consumer of home energy. This is usually done in a standard tank heater that continually heats up a large amount of water. By installing a solar water heater on the roof, nearly all of the hot water you need can be produced for free—without using any gas or electricity.

What will this project do for your home?

A system that uses a solar hot water heater (also referred to as "solar thermal") uses the heat of the sun to preheat water, so a standard heater rarely needs to turn on. Sunlight directly heats water in a collector box on the roof. Once heated, the water travels from the roof back down into the heater tank. Almost half of the homes in the U.S. receive enough sunlight for a solar water heater to work well, and on average, a solar thermal system can provide 50% to 80% of a household's hot water needs. In hotter places such as Florida or Arizona, it could likely supply all of the hot water for a home.

What will this project do for the Earth?

A solar hot water heater uses no energy and creates no emissions or pollution to operate. If half of the houses in the U.S. switched to a solar hot water heater, the savings in carbon emissions would be the same as doubling the fuel efficiency of every car in the country.

Green $pecs

Overall Rating

Difficulty

Green Benefits

Will you need a contractor?

A domestic solar hot water heater is complex, requiring careful installation, connections to existing water lines, and running pipes through walls. Hire an experienced, licensed, and reputable contractor.

Oddly enough, most solar panel contractors (the ones who install photovoltaic panels) do not provide or install solar thermal systems. Although both types of contractors install solar-based systems on your roof, they are as different as electricians and plumbers. You can locate local installers through a directory certified by the North American Board of Certified Energy Practitioners (NABCEP; www.nabcep.org).

The choice of solar water heater is dependent on your climate, sun exposure, and the angle and direction of your roof. An experienced installer will take all of these into consideration to select the appropriate system for your home and budget. Request a written estimate before starting any work and ask if permits will be required in your town. Depending on where you live, you may also need approval from the local planning board or homeowners association.

What are the best sources for materials?

Solar thermal (hot water) is often confused with solar photovoltaic (PV) panels. Solar thermal uses direct transfer of heat from sunlight to warm water. Nothing is converted into electricity, making a solar thermal system highly efficient and much less complicated than a solar PV system. Solar hot water heating systems are available directly from several manufacturers, who may be able to suggest installers in your area. Ideally, solar water heaters are placed on a sunny, south-facing area on the roof directly above the existing hot water tank. A solar water heater takes up less space than solar photovoltaic panels, usually only needing a 24-in. × 48-in. space. The roof needs to be strong enough to bear the weight of the unit, which, when filled with water, can be as much as 500 lb., depending on the system.

If there isn't adequate space on the roof, the box can be mounted in any sunny location within 100 ft. of your home, but keep in mind that ground-mounted installation is more expensive, less efficient, and tends to be more of an eyesore. The height, convenient access, and unobstructed view of the roof make it the preferred spot for the system.

Selecting the correct size for your solar hot water heater will determine if you can get all of your hot water from the sun or just a percentage. You will need about 10 sq. ft. of roof area for each member of your family. A family of four will need around 40 sq. ft. of solar collectors to produce all of their hot water in the summer. This varies slightly based on the specific climate but is a good rule of thumb.

Since the sun powers a solar water heating system and the sun can be unpredictable, a backup system is always required. For extended periods of cloudiness or times when a lot of hot water is needed, a backup system ensures you will always have hot water available. Depending on its age, size, and condition, you may also be able to keep your existing water heater for use as a storage tank.

Solar hot water heating systems consist of two main parts: a storage tank and collector. The storage tank is a modified version of a hot water tank. The collector is usually installed on the roof and available in three main types: batch collectors, flat plate collectors, and evacuated tube collectors.

The simplest solar hot water heating system is called a passive system. Following the principle that heat rises, heated water rises naturally through any type of collector (called "thermosyphoning"). The hot water is stored in a tank where energy can be used to provide additional heating, but no pumps are needed.

An active system uses pumps and requires more energy to operate and costs more to install. A pump may be needed depending on the distance between the storage tank and the collector. The solar heating system can be combined with several other technologies to provide part or all of the hot water for the home.

In freezing climates, a nontoxic and nonfreezing liquid called propylene glycol (the principle ingredient in car antifreeze) is used in the collector instead of water. This is called a closed-loop system, and it works surprisingly well in cold but sunny places.

Batch Collector

Consisting of dark tank inside an insulated box, the batch collector heats and stores large quantities of water until needed. The weight of the batch collector is a problem on older roof structures, and this type of collector is not recommended in freezing climates, as it could burst.

Flat Plate Collector

The most common type of solar thermal system, a flat plate collector, is a thin panel containing copper pipes and covered with a sheet of tempered glass. Cold water enters at the bottom. As the sun heats the plate, the water rises naturally and hot water travels to the top of the collector and back to a storage tank inside the home.

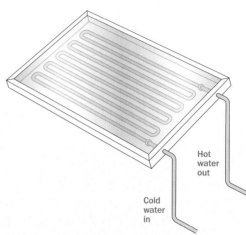

Hot water out

Cold water in

Flat plate collectors typically come in 40-gal. sizes, so households usually invest in two to four collectors to produce all needed hot water. At one-third less weight, they work well on most roof types and in most climates.

How much maintenance will be required after installation?

Since it has no moving parts, a solar thermal system is typically warranted for 20 years and should last anywhere from 15 to 40 years.

Minor periodic maintenance is required, like rinsing the collector with a hose to keep it free of dust, leaves, and debris; checking for cracks or yellowing in the glass top of the collector; and checking that the seals around the pipes penetrating the roof are not cracked or leaking.

Routinely check the pressure-relief valve at the storage tank as well. It should not be stuck in the open or closed position. The pipes should not have any calcium or mineral buildup; call a plumber for any issues with the pipes. If you have "hard" water, you may need to add special water softeners through the collector every three to five years.

How long will the project take to accomplish?

After receiving an estimate, hiring a contractor, and ordering the equipment (it may take several weeks for it to arrive), an experienced

Evacuated Tube Solar Hot Water Collector

The most efficient type of solar heater, an evacuated tube resembles a big thermos with a large tube along the top that feeds into smaller tubes. The space between the tubes is a vacuum, increasing heating efficiency. Although they cost twice as much as flat plates, the evacuated tube collectors work in cloudy and cold locations year-round.

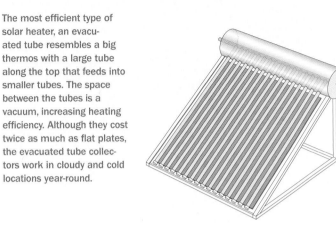

installer should complete the project in about a week. Coordinate ordering the system with the weather; you don't want to install one during the winter or rainy season.

KEVIN SAYS Who doesn't want free hot water? It seems simple, except it's not. Solar heat works great on sunny days. That means cold, cloudy areas of the country won't see as much benefit from the systems, and even bright, sunny areas only get the benefit during daylight hours. The good news is that you can supplement solar-heated water with conventional water heaters and many use tank heaters for storage. No doubt the increased interest in solar thermal is pushing manufacturers to make better products and carry better warranties, but so far manufacturers aren't sharing specifics.

What is the capital cost? $$$

Installation of a solar heating system is pricey, and you also don't save by eliminating the cost of a conventional water heater since nearly all systems require storage tanks and backup systems. A quality solar system will run anywhere from $3,600 to $9,000, with installation, depending on the size of the panels.

RESOURCES

**Solar water
heating systems:**

EnerWorks Inc.
www.enerworks.com

Heliodyne™, Inc.
www.heliodyne.com

**North American Board
of Certified Energy
Practitioners**
www.nabcep.org

Solar Energy, Inc.
www.solarenergy.com

**Taylor Munro
Energy Systems Inc.**
www.taylormunro.com

What financial resources are available? $$$

There are a number of state programs for rebates on solar water systems ranging from $450 to as high as $2,700 based upon usage. The federal government offers a credit for 30% of the project's cost up to $1,500, provided you haven't already used up the credit on other energy-saving projects and the project is done before 2017. Ultimately, you might cover, on average, as much as 20% of the cost.

What is the monthly cost or savings? $

Savings for solar hot water varies based on where and how you live. Hot water usage comes primarily from showers, with laundry and dishwashing contributing a smaller amount to usage. Early morning or late night bathers and those who wash clothes and dishes after dark will see minimal benefit compared with those using hot water during the day. At the low end, you might see only $30 to $40 in annual savings, which means half a century will go by before you earn back the cost of the unit. In the best case, government studies figure about $250 in annual savings, which still means more than 15 years before recouping the cost of the system.

What is the long-term home value? $

Conversations with several appraisers revealed that there is still no data to prove that homes with solar water heating command more in the marketplace. There is interest, but most buyers are concerned with maintenance issues. If the warranty is transferable or if it can be covered under a home warranty, buyer concerns can be resolved. Remodelers, real estate agents, and solar companies maintain that buyers will pay more, but so far there is no firm data to support it.

THE BOTTOM LINE IS...

Solar water heating is one of those green projects that seems like a no-brainer but is actually not that financially attractive when you get down to it. Most families are unlikely to see a ton of savings, and the cost of a system hasn't reached a reasonable point yet. Ten years of technological advancement could bring the cost down, but most likely it will have to be new homebuilders who drive solar water heating into the mainstream. ▮

Install a cool roof

ERIC SAYS The reason we wear lighter colors in hot summer weather is because dark colors absorb heat, making a hot day feel even hotter. The same is true for our homes. A dark roof can increase the temperature at the roof surface by up to 70°F. All of that heat forces air-conditioners to work harder to keep homes cool.

Air-conditioning accounts for more than 15% of all electricity usage in the United States. This is remarkable given that most air-conditioning is used for only a few months of the year. Air-conditioning also contributes to peak electrical loads, straining utility systems and pushing demand for more polluting power plants.

By installing a light or reflective roofing material, called a "cool roof," you can reduce the strain on your air-conditioning system and save energy, all the while making your home more comfortable.

What will this project do for your home?

About $11.3 billion is spent each year to cool homes. Installing a cool roof can cut that bill by up to half and keeps a home comfortable, especially on hot days. By lowering the temperature of the roof, an air-conditioner works less and lasts longer, reducing the amount of maintenance required.

If you live in a place with cold winters, you might think that having a dark roof heating the home is a good thing. Although a dark roof will get hot in the winter, it will not do much to heat the inside of your home. Heat rises, so any heat at the roof remains there. Plus, it takes much more energy to cool a building than to heat it, making it more cost-effective to design for cooling a roof rather than heating it up.

Green $pecs

Overall Rating

Difficulty

Green Benefits

Reflectivity of Common Roofing Materials

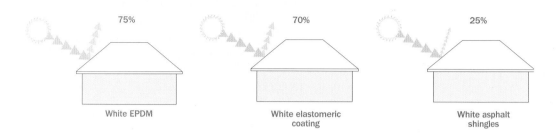

75%
White EPDM

70%
White elastomeric coating

25%
White asphalt shingles

What will this project do for the Earth?

About 64% of homes in the U.S. have air-conditioning, and that number increases each year. Our homes consume a whopping 258 billion kWh of electricity each year to do the job. If we could conserve a quarter of that by installing cool roofing, it would save enough energy to power more than 5.8 million homes for an entire year, and prevent the release of more than 48 million tons of CO_2 into the atmosphere, the equivalent of planting nearly 242 million trees.

Reducing the need for air-conditioning during the peak loading times also reduces the need to build new power plants, which lowers the risk of brownouts and power outages. And lowering the temperature of a roof drops the ambient temperature around a building. Your neighbors would feel a noticeable drop in temperature if everyone around them installed a cool roof. This is referred to as the "heat island effect," and has a cumulative effect on the overall temperature in urban areas. For example, an increase of just 1°F in temperature on a hot summer day in Los Angeles causes a 300-megawatt increase in the city's electric load. Cool roofs can mitigate this effect.

Will you need a contractor?

A roof coating can be painted onto an existing roof to create a reflective surface. The coatings are usually white or silver in color. Applying the coating requires no special skill and can be used on most types of roofing materials. But walking around on a roof isn't for everyone, so use appropriate safety precautions if you decide to take on this project yourself.

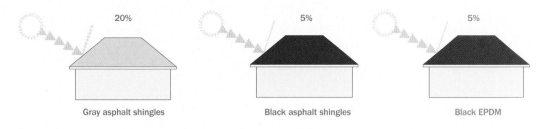

Light-colored roofs reflect more heat than darker materials. Here, the reflectance of standard roofing materials is compared.

20%   5%   5%

Gray asphalt shingles   Black asphalt shingles   Black EPDM

(Adapted from data from Florida Solar Energy Center)

If you're considering replacing your roof (or if you're putting in a new one), all it takes to install a cool roof is to select a nonasphalt roof material in a reflective color. Roofing has to be installed precisely to avoid voiding any warranty and preventing future leaks. Use an experienced and licensed contractor.

What are the best sources for materials?

The reflectivity of a roof is measured by the solar reflectance index (SRI). Select roofing materials with an SRI of 78 or greater for low and flat roofs. For sloped and steep roofs, which do not receive as much solar heat gain, look for an SRI of 29 or greater.

Cool roof coatings are applied like paint, using brushes, rollers, or paint sprayers. Coatings are available as a ceramic-, cement-, or elastomeric-based material. Some coatings can provide additional insulation value as well to the roof.

The key to applying a cool roof coating is preparing the roof properly. Start by lightly spraying the roof with a mixture of bleach and water to kill any mildew. Replace any missing and damaged shingles. Lastly, have the roof pressure-washed to remove any remaining debris. Metal roofing should be free of rust; tiles should not have any loose dirt or grout on them. Allow the roof to dry overnight before applying the cool roof coating.

In an ideal world, asphalt shingles, no matter what color, would not be used with a cool roof because they absorb nearly all of the sunlight hitting them. But the reality is, many of our homes have asphalt shingles. So the solution is to apply cool roof coating as you would to an existing roof. To prevent future issues with moisture, do not apply the coating over damaged asphalt shingles.

DO THE MATH

To further lower the temperature of a roof, install a radiant barrier below the roofing, inside the attic. Radiant barriers are sheets of foil that reflect the heat that comes through the roofing. Air vents along the roof soffits and ridges allow heat to escape from the attic. A solar-powered attic fan can remove any remaining hot air from the attic space (see p. 117).

How much maintenance will be required after installation?

Roof coatings need to be reapplied every five to seven years, but new applications are applied in light coats, requiring only half of the paint of the first application. The coating will also extend the life of the roof beneath it.

Since a cool roof reflects light, it is important to keep the roof surface clean. Periodically check the roof for built-up debris and hose it down with water.

How long will the project take to accomplish?

Applying a cool roof coating can be done in a week. The time it takes to reroof varies based on the complexity and size of the roof but can typically be completed in a few weeks.

KEVIN SAYS Cool roofs are a good idea in warm areas where air-conditioning is a mainstay. Temperate-climate neighborhoods in New England won't necessarily benefit as much. In my house in Connecticut, between my oak trees and sea breezes, I don't have the need for air-conditioning. Therefore, a cool roof for me poses zero energy-saving prospects financially or for the planet. There is also concern out there that cool roofs may cause heat loss in colder months, but most studies show the benefits far outweigh the losses.

What is the capital cost? $$

The cost of reflective roofing materials versus nonreflective options is negligible, so if installing a new roof you can create a cool roof at no additional cost. Those with tar and paper roofs can manage a relatively inexpensive conversion at a price of roughly $0.20 to $0.30 per ft. plus labor. This means a 2,000-sq.-ft. home is likely to cost only $400 to $600 in materials. The cost of reflective paint for shingled roofs is quite a bit more at roughly $1 per ft. (about $2,000 for materials alone for a 2,000-sq.-ft. house). Add on about $500 to $1,500 for labor, depending upon the slope and complexity of the roof. Total cost would be $900 to $3,000.

What financial resources are available? $

Most programs for residential cool roofs only apply to new roof materials rather than coatings, leaving you pretty much on your own to cover the cost of the latter.

What is the monthly cost or savings? $$$

Savings vary with a cool roof. They only apply to air-conditioning and mostly during peak sun times. Various reports state anywhere from a 10% to 20% reduction in air-conditioning costs. So if your average annual expense for air-conditioning is $1,500, you could see as much as $300 (or $25 monthly). More likely you will see somewhere around $10 to $15. At this rate, it will take roughly 3 to 10 years to recoup the cost if you coat the roof yourself.

What is the long-term home value? $

There is no specific data available on the value of a home with a cool roof. As long as the roof is aesthetically pleasing, you may be able to appeal to buyers by showing the difference in energy costs and tout the protective quality of the coating, which may keep them from negotiating for a new roof.

THE BOTTOM LINE is...

Cool roofs *are* pretty cool if you are a big air-conditioning user in a warm, sunny climate. They are relatively low cost and an easy way to make an immediate impact on the planet as well as your wallet. This is one project that should be easy enough to do yourself over a few weekends and will be well worth the time and effort. ∎

RESOURCES

Hawaiian SunGuard
www.hawaiiansunguard.com

National Coatings
www.nationalcoatings.com

Pick a green countertop

Green $pecs

Overall Rating

Difficulty

Green Benefits

ERIC SAYS Granite has emerged as the countertop of choice for many homeowners and is standard issue for new, high-end homes. Unfortunately, of the more than half billion sq. ft. of granite sold each year, much of it comes with a serious hidden cost.

On the surface, granite might appear to be a green material. It's natural, can be salvaged for reuse, and is incredibly durable and maintenance free. But the effort required to mine it leaves behind an incredible amount of damage. Granite is found throughout the world, but to extract it affordably, strip-mining operations cut corners and seek out cheap labor. In China, the government has dropped regulations monitoring radiation in granite slabs, which flooded the market with inexpensive (but radioactive) granite countertops, potentially exposing people to high levels of radiation. In countries such as Nigeria, child labor is often used to cut costs, employing boys as young as four years old who are paid $0.35 a day to quarry the stone. And most granite comes from far away. Its heavy weight requires an incredible amount of fuel to ship to the United States. But the popularity of granite remains, even though we are likely to mine all the granite available in the world during our lifetimes and leave behind empty quarries that leach toxic heavy metals into the soil.

Marble, slate, and sandstone are all common countertop materials that are often generically referred to as granite, and all of them have the same economic and environmental side effects. Alternatives such as solid surfacing are durable with a finished surface that is nontoxic, nonporous, and does not promote the growth of bacteria, mold, or mildew. But the oil-based epoxy resins they are made from consist of a mix of aluminum ore (called bauxite) and acrylic polymers. Dust from cutting these materials can lead to respiratory illness. And the chemical hardeners used, such as the controversial chemical bisphenol A

Countertop Materials: Pros and Cons

COUNTERTOP	BENEFITS	DISADVANTAGES	THINGS TO LOOK FOR
Granite and stone	Durable; heat resistant	Needs to be sealed; stains from water, wine, or lemon juice; joints are visible	Reclaimed and locally produced stone—avoid stone from China and Nigeria
Solid surfacing (Corian®, Avonite™)	Nonporous; stain resistant; does not promote growth of mold or bacteria; no sealing necessary; stains and scratches can be buffed out	Oil-based product; difficult to recycle; dust is bad for respiratory system; uses toxic chemical hardeners	High recycled content; Greenguard certification
Engineered stone (Zodiaq®, CaesarStone, Silestone®)	Nonporous; stain resistant; does not promote growth of mold or bacteria; no sealing necessary; stains and scratches can be buffed out	Overly uniform in appearance; not as durable as stone; oil-based product; difficult to recycle	High recycled content; Greenguard certification
Stainless steel/metal	Durable; hygienic; stainproof	Scratches easily; shows fingerprints; steel has high embodied energy	High recycled content; avoid virgin stainless as it contains chromium
Laminate (Wilsonart®, Formica®)	Low cost; low maintenance; infinite number of patterns	Not heat resistant; uses toxic glues and plastic; scratches easily; difficult to replace or repair	Natural fibers instead of plastic; high recycled content; zero-volatile organic compound (VOC) glues used
Ceramic tile	Wide range of types and costs available; heat resistant; single tiles can be replaced; creates less waste	Shows grout joints between tiles; hard to clean grout; tiles may chip; bacteria can grow in joints	High recycled content
Recycled glass terrazzo (Vetrazzo®, IceStone®, EnviroGLAS®)	Infinite number of color combinations; durable, heat resistant; high recycled content; attractive	May require sealing; difficult to locate fabricators	Only 100% recycled glass, preferably from local sources
Paper composite (Richlite®, PaperStone™, EcoTop™)	Can incorporate drainboards or trivets; high recycled content; can be used for cutting	Limited number of colors; scratches show easily	Forest Stewardship Council-certified or recycled paper only
Wood/butcher block	Natural; can be used for cutting; can be repaired	Not good around wet areas; not heat or scratch resistant; usually only good for a portion of the counter	FSC-certified wood; reclaimed wood
Concrete	Great for curves; can incorporate drainboards or trivets; durable; heat resistant; do-it-yourself kits available	Must be sealed; porous; scratchable; cement has high embodied energy	Substitute up to half of the cement with fly ash
Recycled plastic (Yemm & Hart)	Recycled material; can be cut and molded easily; thousands of colors; great around wet areas/sinks	Easily scratches; feels like plastic; nonbiodegradable	100% recycled plastic
Glass tiles (Blazestone, UltraGlas®)	Easy to clean; high recycled content; nonporous; heat resistant	Shows grout joints between tiles; hard to clean grout; tiles may chip; bacteria can grow in joints	Should use at least 70% recycled glass

(BPA), are known endocrine disrupters linked to dangerous health side effects, including reproductive and developmental defects in children.

Luckily, dozens of green, healthy, and attractive alternatives are available that make it easy to give up granite countertops and the like.

What will this project do for your home?

Lax regulations on the export of granite have flooded the market with cheap stone containing high levels of radon radiation. Some radioactivity is natural for any material mined out of the earth, but homeowners are discovering much higher levels than normal. It can increase the risk of cancer about as much as smoking half a pack of cigarettes a day.

Selecting a green countertop brings a healthy material into your home. In choosing an alternative countertop, your kitchen will stand out from the rest.

What will this project do for the Earth?

Approximately 500 million sq. ft. of granite are installed every year, with the number increasing slightly every year. Extracting the stone comes with countries forcing children into slave labor and others avoiding regulation compliance. By avoiding granite and stone countertops, you directly reduce demand for inferior and sometimes harmful materials and unscrupulous practices.

Will you need a contractor?

Installing a countertop can be tricky. Find local and experienced cabinet installers, kitchen contractors, and general contractors to help with this project. If using a kitchen designer, you may want to use one certified by the National Kitchen and Bath Association (NKBA; www.nkba.org). Its website provides a free directory of certified professionals.

What are the best sources for materials?

It is easy to find green, healthy, and less impactful countertop materials if you know what to look for. Research countertop materials online, but don't make a decision until you see samples in person. If the manufacturer charges for samples, offer to return them when done. If they don't want them back, donate them to your local architecture or interiors school. When looking at various materials, ask about the consistency of colors and how the final appearance may differ from the sample.

How much maintenance will be required after installation?

There is no perfect countertop material. They all need different levels of maintenance and care, so make sure to research the pros and cons of any material (see chart on p. 159) and ask the manufacturer for care instructions so you know what you are in for once the countertop is installed.

How long will the project take to accomplish?

Replacing countertops requires a few steps. The materials must be ordered, which may take up to eight weeks to arrive if not in stock. Then the fabricator must shape and cut the countertop to fit your kitchen. This can take another two weeks. Finally, the finished counter is installed over cabinets, a process that can take several hours.

Recycled material countertops, like this one made from recycled glass set into grout, provide durable, attractive, and eco-friendly alternatives to granite.

KEVIN SAYS Manufacturers are well aware of the "wow" factor that comes with countertops. It's one of the best places to show personality and charm in the home. Replacing counters just for looks, however, can do more harm to the environment than leaving well enough alone, creating waste and burning fossil fuels for transportation of new counters. But if you've decided to update your kitchen, by all means spice it up with an eco-friendly countertop choice that is just as affordable and attractive as those that are not so kind to the planet.

What is the capital cost? $$$

Countertop prices range anywhere from $45 per lin. ft. for Formica to $250+ per lin. ft. for exotic stone. There is a broad array of stone and eco-friendly materials available in the $60 to $150 range, including installation and backsplash. A typical kitchen may need from 30 lin. ft. to 50 lin. ft., with costs on the low end ranging from $1,800 to $3,000 and going up to $4,500 to $7,500.

Although extremely popular, most solid-surface countertops are made from chemical epoxy and plastics. Healthier alternatives are composite materials, such as this one from recycled paper, which can be shaped and edged to accommodate any style.

What financial resources are available? $

Because countertops aren't considered an energy-saving item in and of themselves, governments and utilities aren't particularly interested in rewarding a green material choice over a nongreen one. If you are buying a new home, builders often offer big purchase incentives such as upgraded countertops, but they generally don't discriminate between standard and eco-friendly choices.

What is the monthly cost or savings? $

Once countertops are installed, there should be little or no monthly maintenance, so there is no money to be saved here, but it shouldn't cost you anything either.

What is the long-term home value? $$$$

There is no question that the kitchen is definitely a place where investment can pay off in home equity. But rarely do you see a 1-for-1 return on investment for remodeling. The key is how it looks. Virtually no data exists to support the belief that people will pay more for an eco-friendly

Up to half of the Portland cement in typical concrete countertops can be replaced with fly ash, which reduces the impact of concrete on the environment.

countertop than for granite. People will, however, pay more for nice-looking counters than for old, drab ones. If the kitchen is beautiful and well put together, then you could recoup your cost in the midrange of $60 to $100 per ft., provided the neighborhood prices support having a gourmet kitchen.

THE BOTTOM LINE IS...

Countertops are a great way to dress up a fading kitchen. If your kitchen is in need, by all means pick something green and appealing. Certainly there is no need to overspend since the range of eco-friendly options can deliver for most budgets. But if your existing counters look nice and are in good shape, you'll create more environmental damage by replacing them than keeping them as is. Remember the old adage: If it ain't broke, don't fix it. ▪

RESOURCES

Blazestone
www.bedrockindustries.com

BottleStone™
www.bottlestone.com

UltraGlas
www.ultraglas.com

Wilsonart
www.wilsonart.com

Yemm & Hart
www.yemmhart.com

Cover walls with healthy finishes

Green $pecs

Overall Rating

Difficulty

Green Benefits

ERIC SAYS We spend more than 80% of our time indoors. Breathing in stale, recycled air from a heater or air-conditioner takes a toll on our health, and a lack of fresh air is worsened by noxious chemicals in our homes. Asthma rates have tripled in adults since 1980. In children, asthma has gone from the seventh-leading chronic illness to the first in less than 20 years. Poor indoor air quality is largely responsible for this increase, and the finishes on our walls play a big part.

We fill our homes with numerous products that release VOCs into the air—harmful chemicals linked to cancer and other diseases. The primary source of VOCs is interior paint.

Although most paint manufacturers produce low-VOC paints, they still contain VOCs. Consider using earthen plasters instead. With natural plaster finishes, you eliminate the primary source of VOCs in the home and update the look of the walls in your home at the same time.

What will this project do for your home?

By selecting earthen plasters, you can eliminate that new paint smell once and for all. Imagine having no odor right after painting! Earthen plasters do not save energy or water but definitely contribute to the health and safety of everyone in your home.

What will this project do for the Earth?

By choosing healthier finishes, unneeded chemicals and toxins are prevented from entering our air, land, and water. Making the simple change to a healthy plaster will reduce some of the 69 million gal. of harmful chemical paints that end up in our nation's landfills each year.

Will you need a contractor?

For a smooth plaster finish, you will need a professional to apply it to a large area. If you are using the plaster on just a single wall, you could try it yourself. The plaster is applied with a trowel. It is fun to apply, but it takes practice to achieve a consistent finish.

What are the best sources for materials?

Earthen plasters are made from recycled and reclaimed aggregates, natural clays, and pigments. They contain zero VOCs and give walls pattern and depth. They can be troweled on or sprayed on using a drywall hopper over most types of wall surfaces. Extra preparations need to be made to the wall if you're installing it over tile, brick, or stone.

Another healthy option, lime-based paint is simple whitewash, consisting of lime and water. Lime paint goes onto the wall more like watery plaster than paint. The finished surface resembles plaster.

If a home was constructed before 1978, it may contain traces of lead paint. If you suspect lead paint in your home, contact a professional abatement specialist to remove the flaking paint before repainting.

Never dump paint or plaster down the drain. It pollutes the local watershed. Plaster can be stored after it has been mixed and left to dry out. It only needs to be broken up and reconstituted for touchups or a new application.

How much maintenance will be required after installation?

Plastered walls can be kept clean with a damp sponge. Chips in plasters can be patched with varied success.

How long will the project take to accomplish?

Plastering takes about twice as long as painting. If you hire a professional, any additional time required will be unnoticeable. The drying time for plaster is less than for regular paint.

Earthen plaster is an attractive, healthy, and durable alternative to paint. It is made with natural clay and pigments as well as recycled and reclaimed aggregates such as sand and shell.

KEVIN SAYS Earthen plasters are limited in colors, which is not surprising since the name itself implies earth tones. They have a nice look but come with moisture issues and can crumble when bumped. Luckily, they can be repaired, but they may not be a practical option in a home with kids or rambunctious pets. Sealers are an option, but they change the look of the plaster and some have the very chemicals we are trying to avoid. If you are desperate for that textured look, lime plaster is a more durable albeit costly option.

What is the capital cost? $$

Earthen plasters can be a nice decorative touch for homes but are certainly financially beyond the necessity for planet-saving home décor.

Sources of Poor Indoor Air Quality

CONTAMINANTS	SOURCES
Volatile organic compounds (VOCs)	Furniture polish, cleaning solvents, pesticides, carpet dyes and fibers, glues, adhesives, sealants, paints, stains, varnishes, strippers, wood preservatives, air fresheners, plastics
Formaldehyde	Particleboard, interior-grade plywood, cabinetry, furniture, urea formaldehyde foam insulation, carpet, fabrics
Pesticides	Insecticides (including termiticides), rodenticides, fungicides, disinfectants, herbicides (from outdoor use)
Lead	Lead-based paint, exterior dust and soil
Carbon monoxide, carbon dioxide, nitrogen dioxide	Improperly operating gas or oil furnace/hot water heater, fireplace, woodstove, unvented gas heater/kerosene heater
Sulfur dioxide	Combustion of sulfur-containing fuels (primarily kerosene heaters)
Respirable particulates (RSPs)	Fireplace, woodstove, unvented gas or kerosene heater
Polycyclic aromatic hydrocarbons (PAHs)	Fireplace, woodstove, unvented kerosene heater
Biological contaminants	Plants, animals, humans, pillows, bedding, house dust, wet or damp materials, standing water, humidifiers, evaporative coolers, hot water tank
Asbestos	Pipe and furnace insulation, ceiling and floor tiles, shingles and siding
Radon	Soil and rock, some building materials, water

Compared with painting, earthen plaster costs five times as much, or even more if you use a lime plaster for durability—about $0.80 to $1.35 per sq. ft. plus shipping and application. That means plastering a 12-ft. × 12-ft. room with 9-ft. ceilings would cost roughly $500 per room. You could paint the room yourself for less than $100 with zero-VOC paint. Lime plaster would be nearly $650, and both plaster options become very costly when factoring labor.

What financial resources are available? $

Other than closeout discounts from the manufacturers, I couldn't find much in the way of rebates or credits for earthen plasters.

What is the monthly cost or savings? $

Earthen plasters may require maintenance, which is fine if you have the time and skill to repair them. Otherwise you may find yourself hiring someone to patch walls periodically.

What is the long-term home value? $

Earthen plasters may impact the value in higher-end homes. Lime plasters carry a little more distinction and, installed properly, have a nice visual appeal. But these wall finishes usually involve regular maintenance, which may detract some buyers.

THE BOTTOM LINE is...

Earthen plasters are a beautiful option for those who are comfortable living in a high-maintenance environment. I personally prefer to be the high-maintenance subject in my home. ■

RESOURCES

Earthen plaster:

American Clay
www.americanclay.com

Lime-based paint:

TransMineral USA
www.limes.us

Install healthy wall coverings

Green $pecs

Overall Rating

Difficulty

Green Benefits

ERIC SAYS Wallpaper and wall coverings are an $80 million-a-year business. From stripes to polka dots, wall coverings come in every imaginable pattern, color, and texture, and wallpaper adds character and interest to a room. But paper wall coverings are printed using oil-based inks on virgin wood paper sources. At a time when 96% of old-growth forests are gone, creating millions of square feet of wallpaper from trees seems off-base, and printing with ink made from oil is ill-advised, given we import 70% of our oil from foreign sources and the price continues to climb.

Commercial-grade wall coverings are not any better. They are washable, durable, and mold resistant, but are typically made from vinyl (called PVC, or polyvinyl chloride). Vinyl is a significant source of volatile organic compounds (VOCs) and releases toxins (namely dioxin) when thrown away. It's also highly toxic to produce, difficult to recycle, and releases a deadly smoke (containing hydrochloric acid) when burned. The chemicals used in vinyl (called phthalates) also disrupt the human endocrine system, especially in children.

In addition, most standard wall coverings do not allow moisture to pass through them, so walls cannot breathe, trapping condensation behind the wallpaper, which can lead to mold. And many wall coverings are treated with antimicrobial and stain-resistant additives that can release harmful chemicals into indoor air.

Smarter and healthier options, made with recycled paper, natural fibers, and plant-derived inks, are now readily available for an incredible assortment of wall coverings, so you can beautify your walls without sacrificing your health or the environment.

What will this project do for your home?

Choosing healthier wall coverings containing no PVC or chemical additives greatly improves the air quality in a home. Breathable fabrics and papers reduce the risk of mold in walls.

What will this project do for the Earth?

More than 5 million yards of wall covering are discarded worldwide each year, which are then burned or end up in landfills, slowly leaching toxins into the air, ground, and water. By switching to healthy, non-PVC wallpapers and adhesives, the amount of pollutants dumped into the environment is reduced.

Not Just Any Wallpaper

Look for wall coverings that have these characteristics:

- **Non-PVC or PVC free**
- **Water-based coating**
- **Free of plasticizers containing phthalates**
- **Free of elemental chlorine in backing and inks**
- **Free of heavy metals, especially in inks**
- **Free of mercury and cadmium**
- **Free of ozone-depleting chemicals**
- **Free of arsenate in the antimicrobial coating**

There are literally hundreds of brands of healthy wall coverings available that meet these specifications. Green Home Guide (www.greenhomeguide.com) and Greenguard (www.greenguard.org) provide free directories of healthy wallpapers on their websites.

Will you need a contractor?

Installing wallpaper can be a fun project to do on your own, although it requires lots of stretching, reaching, and climbing on a ladder. Don't overexert yourself, and call in the professionals if the project seems too much to handle alone.

What are the best sources for materials?

The funny thing about wallpaper is that most of it is not paper at all. Healthy wall coverings are available in many natural materials, such as latex, grass, straw, sisal, cellulose, honeysuckle, cork, flax, silk, linen, and mulberry, as well as recycled materials, including paper, polyester, and cardboard. Some are made with water-based vegetable dyes, while others have self-adhesive backing, eliminating the need for paste. Look for companies using natural materials, high recycled content, and low chemical manufacturing.

Installing wallpaper requires a tremendous amount of adhesive, and not all wallpaper pastes are the same. Vinyl, for example, requires a different adhesive than those for fabric or paper. Check with the manufacturer for the proper type of wallpaper paste.

ABOVE If selecting a fabric wall covering, choose one grown without pesticides or synthetic fertilizers. Hemp, flax, and linen require less energy to produce than cotton.

RIGHT Don't be fooled by manufacturers calling a product that contains just 5% recycled content green, when 80% and above is available. This wallpaper from MioCulture is made from 100% recycled paper.

If installing a porous material, like sisal or burlap, expect to use up to twice as much paste. Look for natural, plant-based adhesives (see Resources on p. 172). Some pastes are premixed; others are sold in a powder form that must be mixed with water.

How much maintenance will be required after installation?

When installed over a properly prepared surface, wallpaper should provide a durable and low-maintenance finish.

If you live in a hot, humid climate, the outside air can form condensation between the back of the wallpaper and the wall. Look for the wallpaper's perm (permeability) rating as a measure of how it breathes. If you live in a humid place, choose a wall covering with a perm rating of at least 13 or higher. The higher the number, the more breathable it is.

How long will the project take to accomplish?

Wallpapering a room can usually be done in as little as two days, including preparing the wall. The more time you spend planning and prepping, the less time the project will take in the long run.

KEVIN SAYS There isn't any functional need for wall coverings in a home, but when done right they can cause people to ooh and ahh. But done wrong, they can be downright deadly. Well, sort of—there are theories that Napoleon Bonaparte's wallpaper did him in. In the early 19th century, certain pigments in wallpaper patterns contained high levels of arsenic. Bacteria growing in flour-based wallpaper paste reacted with the arsenic and evaporated into the air, which would explain why Bonaparte's autopsy revealed toxic levels of arsenic in his body. Luckily, today's wall coverings don't have arsenic, so that risk is gone and high style can be enjoyed at a different price. So if you have decided to make the plunge and invest in wall coverings, there are plenty of eco-friendly coverings at comparable cost to nongreen options.

Homemade Wallpaper Paste

One of the healthiest ways to put up a new wall covering is with your own wallpaper paste. This formula comes from an old recipe and works just as well as any commercial paste on the market.

1. Mix 1 cup flour (wheat, corn, or rice) with 3 teaspoons alum (available in the herb section of supermarkets), and place in a double boiler (or place a smaller pan inside a larger one of boiling water).
2. Add water until the mix has the consistency of heavy cream.
3. Gently heat until the paste has thickened to resemble gravy.
4. Allow it to cool.
5. Stir in 10 drops oil of cloves (available from pharmacies and health-food stores).
6. Pour into a glass mason jar with a screw top.

This mix makes 1 cup of paste and, thanks to the oil of cloves, it has a shelf life of two weeks when kept refrigerated.

What is the capital cost? $$$

The good thing about eco-friendly wall coverings is that there are a variety to fit many budgets. Of course, how much you spend will depend on how much you use. Fabrics run about 54 in. wide, or a little over 4 ft., and wallpapers run a little over 2 ft. wide. That means 1 yd. of fabric covers about 12 sq. ft. and 1 yd. of paper covers about 6 sq. ft. To cover 850 sq. ft. to 1,000 sq. ft. in fabric requires about 85 yd., and for paper, about 150 yd.

Eco-friendly fabric and paper coverings range in price from $18 to $60 per yd., with really fancy stuff running well over $100. That means covering 850 sq. ft. to 1,000 sq. ft. could cost $1,500 to $5,000 for fabric coverings and as much as $2,700 to $9,000 for paper. And then there is the labor cost. Paperhanger charges vary, but figure $50 to $75 per hour. Best to do it yourself so you can save on that, and cook up Eric's special paste potion (at right).

What financial resources are available? $$

Just because the government won't pay to make your home look pretty doesn't mean others won't help. Many manufacturers and home improvement stores blow out last year's designs at discount prices, with sales on eco-friendly papers ranging from 10% to as much as 50% off.

What is the monthly cost or savings? $

Although there is no inherent monthly cost in choosing wallpaper over paint, there is the question of durability. If going with good-quality eco-friendly wallpaper, it will likely hold up far longer than a coat of paint. That means saving a little on repainting down the line, at least for the walls that are covered. If you pick a trendy wall covering that could go out of style after a few years, you may end up spending more money to replace it.

What is the long-term home value? $$

There is no question that a well-designed home earns a premium price on the market. But if the home showcases a major style faux-pas, it could sit on the market for a while. So just make sure that whatever wall covering goes up is easy on the eyes. Do it right, and you might get back a bit of the dough you invest. If not, be prepared to change it out if you decide to put the house on the market.

THE BOTTOM LINE IS...

Anyone who has had to scrape off old wallpaper knows the durability benefits of wall coverings. There are plenty of stylish and interesting eco-friendly offerings that cost about the same as the unhealthy options. Yes, it's a lot more expensive to use wall coverings than paint, but if you crave that high-fashion look, it might be worth the money. You can always focus on one or two rooms to keep costs down. ▉

Replace vinyl with linoleum flooring

ERIC SAYS Talking about linoleum may cause you to think back to your grandmother's kitchen or elementary school classroom. Linoleum has a long history in our homes and buildings and was the leading choice for flooring in houses for a hundred years, ever since it was invented in 1860.

But with technological advancements came less expensive materials, and by the 1960s, vinyl flooring replaced linoleum in popularity. Today, vinyl is the leading flooring material, with more than 5.8 billion sq. ft. sold each year. It makes up nearly half of all hard-surface flooring sold in the U.S. And although many people mistake vinyl for linoleum and vice versa, the two materials couldn't be more different.

On the surface, vinyl appears to be a great material. It's incredibly inexpensive, durable, and used in thousands of products. Unfortunately, vinyl (also called vinyl composition tile [VCT] or polyvinyl chloride [PVC]) is a health hazard in every stage of its life, from production through disposal. Producing vinyl releases deadly toxins and, although it is technically recyclable, few recycling centers accept it. In the landfill, vinyl leaches dioxin into soil and water, and in a fire, vinyl releases lethal smoke containing hydrochloric acid. No wonder vinyl is often referred to as the "poison plastic."

Made from linseed oil, jute, pine rosin, and pine flour (sawdust), linoleum is all-natural, biodegradable, and antibacterial. It is a thin and soft material but surprisingly durable, and it's available in a wide array of colors. By replacing vinyl floors with natural linoleum, you'll reduce the amount of pollutants and health hazards brought into your home without sacrificing durability and affordability.

Green $pecs

Overall Rating

Difficulty

Green Benefits

Although commonly confused with linoleum, vinyl is less expensive, less durable, and tends to shrink and crack over time. Vinyl goes by many names, including PVC and VCT. Don't be fooled by salesmen trying to sell vinyl by calling it linoleum.

What will this project do for your home?

Linoleum provides a soft walking surface and lasts for up to 50 years. It is naturally hypoallergenic and free of volatile organic compounds (VOCs), so it's completely healthy for you and your family. The surface of linoleum is also antistatic, so it repels dust and dirt that trigger asthma and allergies, keeping a house cleaner.

What will this project do for the Earth?

By replacing vinyl with linoleum—a sustainably harvested material—you'll reduce some of the 14 billion lb. of PVC produced every year in the U.S. and prevent millions of pounds of pollutants from being released into the environment.

Will you need a contractor?

Replacing and installing floor tiles is not difficult, but it can be backbreaking. Square rooms are easy to do yourself, but more complex areas might best be left to a professional. Your local flooring showroom will have a list of recommended installers. Check references and get an estimate before starting any work.

What are the best sources for materials?

Linoleum is an ideal flooring choice for kitchens and bathrooms, pretty much anywhere you would consider installing tile. It is water resistant and easy to clean but (like all resilient tile) not recommended for damp locations, such as basements. Linoleum is ideal for use with radiant heating. The heat will not affect the linoleum, which will quickly adjust to the temperature. Linoleum has a unique smell (from the linseed oil), so sniff a sample in the store if you are chemically sensitive. If using adhesives to install, only select ones with low VOCs.

Linoleum comes in individual tiles (typically 12 in. × 12 in., although other sizes are available) or in large sheets. Sheet linoleum is more cumbersome to install but provides a seamless look when complete. Intricate stencils, borders, and patterns can be cut into the sheets to create stylish design effects. Tiles are better for large rooms, and individual tiles can be easily replaced. Linoleum is available with either a preapplied topcoat or unfinished, and it can be installed over wood, concrete, and most subfloors. The only restriction is that the subfloor must be smooth and level.

When ordering linoleum, measure the rooms to be tiled and add 10% to the square footage to allow for for cut tiles. Once the linoleum arrives, it may be slightly yellower in color than the samples you reviewed. This yellowish film is normal and quickly fades in the sunlight. You may want to wait a few days after installation before laying any area rugs over the linoleum to allow the sunlight to do its magic.

When it comes time to remove old tiles and put down new, be careful. Tiles installed prior to the 1970s may be asbestos tiles in disguise. Asbestos tiles (also referred to as vinyl asbestos tiles or VATs) contained asphalt and usually came in dark colors. They look similar to vinyl or linoleum, but asbestos fibers are linked to lung cancer and were discontinued in the 1970s. Asbestos is only harmful if the material is broken apart. If you have any asbestos tiles in your home, do not touch them! Never sand asbestos or do anything to disturb it. Contact a local asbestos abatement company to do a proper inspection. Unbroken and undamaged asbestos can remain safely in your home if covered completely by a new floor. Old vinyl tiles can simply be removed with a pry bar or large putty knife. Stubborn tiles are removed easily with a heat gun. Once the old vinyl tiles are pulled up, check with your local waste disposal company about recycling them.

Depending on the brand of linoleum, tiles can be installed by spreading adhesive on the floor, by using self-adhesive tiles, or by simply clicking them into place.

There is a misconception that vinyl and linoleum are the same, but linoleum, like the large sheet being installed here, is a natural, biodegradable material made of linseed oil. Vinyl is an oil-based plastic that releases toxic chemicals over its entire life.

How much maintenance will be required after installation?

Instead of wearing away, linoleum hardens in high-traffic areas. The color and pattern are dyed all of the way through the material, so patterns won't fade or wear over time. Prefinished linoleum is virtually maintenance free. Unfinished linoleum will need a coat of water-based sealer annually. With regular cleaning and sealing, a linoleum floor will last for 50 years.

How long will the project take to accomplish?

The time it takes to remove old floor tiles and prepare the subfloor will vary based on their condition. Once the subfloor is clear and ready, covering a floor can be done in several hours. Tiles install slightly faster than large sheets. Allow the adhesive to dry for three days before moving the furniture back and resuming normal foot traffic in the room.

KEVIN SAYS The old complaint about linoleum was that it was drab and boring. The good news is that linoleum is becoming so popular due to its eco-friendly properties that it is available in many more colors and patterns. One of the best things about linoleum is that, unlike vinyl, the color goes all the way through, which means that even when it wears, it still looks rich and new. How funny that ultimately we are revisiting our grandparents' linoleum just to find out that it is as good if not better than the materials that have come about in the meantime. Although the cost of linoleum comes in on the high end of the scale, it also comes with about five decades of durability.

What is the capital cost? $$

There's no question linoleum is more expensive than vinyl options. The least expensive flooring is vinyl squares, ranging from $4 to $8 per sq. yd. This makes the material cost for a 10-ft. × 15-ft. kitchen (needing roughly 17 sq. yd. of material) about $68 to $136. Next—at a cost of $3.50 to $26 per sq. yd. (for really fancy stuff)—vinyl sheeting comes in at $60 to $442 for materials. Linoleum costs the most at $33 to $54 per sq. yd., totaling $561 to $918 for materials. Installation costs

$400 to $600, with linoleum being on the high side. To enhance the appearance of the floor, you might also figure another $250 for coving.

What financial resources are available? $

No specific tax credits or rebates are available for changing out floors. Big-box stores often have flooring sales, but most don't carry eco-friendly flooring just yet.

What is the monthly cost or savings? $

Durability is the primary factor related to the monthly savings of linoleum flooring. Maintenance costs are equivalent across the board, but since linoleum lasts arguably twice as long as vinyl, you can figure on making up the difference within 20 to 25 years.

What is the long-term home value? $$

A well-appointed kitchen always increases a home's value, but no appraiser we could find gives additional value to linoleum over vinyl. In fact, stone or tile generally has the greatest positive impact but is much pricier than linoleum or vinyl. But since the relative additional cost of upgrading from vinyl to linoleum is only around $500, you might get a good real estate agent to up-sell a buyer on the long-term durability. If the buyer likes the look, that is.

THE BOTTOM LINE IS...

Any affordable alternative to PVC is a good choice. Linoleum is a natural, durable, and healthy option for flooring that won't break the bank. Of course, taking out existing vinyl flooring doesn't help our landfill problem, but you'll be set for another 50 years. ■

RESOURCES

Armstrong®
www.armstrong.com

Forbo Marmoleum®
www.themarmoleumstore.com

Install cork, palm wood, or bamboo flooring

Green $pecs

Overall Rating

Difficulty

Green Benefits

ERIC SAYS Forests are the lungs of the planet, filtering air and converting carbon dioxide into oxygen. But more than 80% of Earth's ancient forests have been damaged or destroyed by mankind's activities. According to Greenpeace, an area of forest the size of a football field is chopped down every two seconds. In total, 32 million acres—that's an area the size of Louisiana—of natural forest are logged each year. And what remains is being threatened by illegal logging practices. Every forest on the planet will be gone by the end of the century if this continues.

The majority of logging performed is illegal, exploiting workers and losing revenue for the economy. Associated with unsafe working conditions and child labor, illegal logging accounts for as much as 30% of all hardwood lumber produced. By selecting sustainably harvested material instead of wood—like cork, bamboo, or palm wood—you'll help slow the effects of deforestation and reduce some of the nearly 1.7 billion bd. ft. of hardwood produced each year.

By definition, a sustainably harvested material is one that is gathered without killing or destroying the original source. Rather than chopping down an entire tree to get to the wood, a sustainably harvested material is collected by trimming the plant. Nothing dies in this process, ensuring material for future generations.

What will this project do for your home?

There are several alternatives to wood flooring that can add signature warmth and beauty to a home. Cork, harvested from recycled wine corks or from the bark of a tree, grows back in five to seven years and provides a cushioned, healthy, and mold-resistant floor. Bamboo is a fast-growing plant that can be trimmed without killing it. Like bamboo, palm wood grows quickly and produces a strong wood. These floors are also typically finished with healthier, water-based finishes.

What will this project do for the Earth?

Destroying the Earth's forests has serious consequences. Trees absorb carbon dioxide, converting it into oxygen for us to breathe while slowing the effects of global warming. They help stabilize the weather and climate by pumping humidity into the air.

When a tree is cut down, all of the greenhouse gases trapped in the trunk, roots, and leaves are released into the atmosphere, increasing the effects of global warming. Nearly one-quarter of all carbon emissions come from deforestation, more than all of the emissions from automobiles in the U.S. In the Congo alone, deforestation will release 34.4 billion tons of carbon dioxide by 2050. This is about the same amount as the total carbon dioxide emissions for the United Kingdom since 1950. Choosing sustainably harvested materials for your floors helps slow these effects and reduce global warming.

Will you need a contractor?

Whether you should install a new floor yourself largely depends on the condition of the subfloor. The subfloor is the structural floor below the floor being replaced. It is typically made from sheets of plywood and should be level and undamaged. If you're building a new home or your home has a subfloor in good condition, you could easily install a new floor yourself, but be prepared for several days of hard work.

If you have a concrete subfloor or a subfloor that needs to be replaced, call a professional. Your local flooring showroom will have a list of qualified installers (even if you don't buy the flooring from them).

What are the best sources for materials?

All sustainable flooring types can be installed in any location you would typically install standard wood flooring. Damp locations, such as basements, should be avoided. Also check into installing your green flooring choice as a "floating floor," where the materials click or lock together, avoiding the need for glue or nails entirely.

Cork Cork flooring provides a soft, warm-looking, natural surface that mimics the texture of trees and gives off a scent similar to smoked hickory. Famed architect Frank Lloyd Wright often installed cork in his kitchens due to its natural resistance to water. Originally, cork was

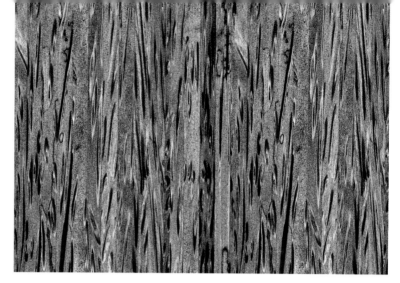

Palm wood, harvested from coconut trees, has a distinctive checked pattern in its grain.

harvested from the bark of the oak cork tree, but cork floors made from recycled wine bottle corks are now available.

Cork floors are sold in standard-size tiles, usually 12 in. × 12 in. or 24 in. × 24 in. Most tiles are prefinished with a light, glossy surface, which is the best option for a durable finish. Avoid cork flooring with a vinyl backing, and look for a zero-VOC adhesive to install the tiles.

Bamboo Bamboo is a fast-growing grass; some species grow several feet in a single day. Stalks of bamboo are cut and split into long strips, which are then glued together to create an incredibly strong floor. Bamboo is often associated with green buildings as the flooring of choice, but the reality of bamboo is slightly more complicated. Most bamboo is harvested in Asia, requiring a great deal of energy to ship it to the U.S., and the popularity of bamboo has also given rise to unscrupulous practices of clear-cutting and destroying bamboo crops to harvest all of it at once, killing the plants in the process.

Bamboo floors are sold in long planks, similar to wood flooring, and install in the exact same way. Prefinished planks are the easiest option as there is no need for a topcoat to be applied later on. Make sure any prefinished sealer is water-based, and if the floor needs to be finished, try to find a zero-VOC finish. Be aware that more and more manu-facturers are adding bamboo to their lines of flooring, and not all are created alike. Look for bamboo products from renewable sources with zero-VOC coatings.

Palm wood and other alternatives Palm wood is harvested from coconut trees, making it a by-product of coconut farming. Like bamboo, palm wood is sold in strips and installs just like a wood floor. Unlike bamboo, however, palm wood comes from cutting down older coconut trees that no longer bear fruit. Palm wood has an unusual checked surface slightly reminiscent of a coconut shell.

Other greener alternatives to traditional wood flooring include reclaimed wood (see p. 226) and wood (and bamboo and palm wood) from forests and lumber mills meeting the strict requirements of the Forest Stewardship Council (see p. 235).

Natural materials vary from batch to batch. The final color of your floor may be slightly different from samples in the showroom. Ask to see a range of samples to get an idea of how much the color can vary.

How much maintenance will be required after installation?

Cork, bamboo, and palm wood are all relatively maintenance free as long as they are kept clean and dry. As with any wood floor, sharp heels or heavy furniture may leave dents in the surface of the floor. Cork floors need recoating with natural carnauba wax twice a year to keep their sheen.

How long will the project take to accomplish?

Once an old floor has been removed, new floors can be installed in a day or two. Before installation, figure out the layout of the materials. Remember to call a local salvage company to come and recycle the old floor. It may have another life after you are done using it.

KEVIN SAYS Eco-friendly bamboo, palm wood, and cork floors are certainly worth a look, but not everything about them gets you into green heaven. Cork is in the clear, but there is debate over bamboo growing and harvesting practices that lead to the clearing of old-growth forests and allow for manufacturing processes that include unfair labor standards and the use of harmful chemicals. If going with bamboo, shop around for a manufacturer that is eco- and labor-friendly.

What is the capital cost? $$$

Good news for fans of newer planet-friendly materials: According to several contractors, installation runs roughly equal for hardwood, bamboo, palm wood, and cork flooring, all coming in at about $1.50 per ft.

RESOURCES

Cork:

Expanko®
www.expanko.com

Habitus
www.habitusnyc.com

Nova Distinctive Floors™
www.novafloorings.com

USFloors®
www.naturalcork.com

Bamboo:

Teragren®
www.teragren.com

Palm wood:

EcoTimber®
www.ecotimber.com

Cali Bamboo®
www.calibamboo.com

Smith & Fong
www.durapalm.com

Cork comes in the cheapest, ranging from $2.90 to $4 per sq. ft. plus installation. Bamboo runs a little more at $3 to $6 per sq. ft. plus installation. Palm is substantially higher at around $8 per ft. In comparison, nongreen hardwood is not going to save you any money. The lowest hardwood flooring comes in around $4 to $20 per sq. ft.

What financial resources are available? $$

The government isn't helping pay for flooring, but the popularity of eco-friendly flooring has dropped its cost. More manufacturers also means more overstock, and a quick look online can yield sales and discounts on last year's styles. So by all means, save and look like last year.

What is the monthly cost or savings? $

Eco-friendly options are no more durable than hardwoods; they are susceptible to damage and, in the case of cork, may need maintenance. But all in all, you will probably pay less than $100 a year on repair and maintenance.

What is the long-term home value? $$

None of the appraisers we spoke with give additional value for cork or bamboo over any other hardwood flooring, but the choice of hardwood flooring itself adds value over a comparably carpeted house when done tastefully. Appraisers were reluctant to give full value for the improvements, estimating instead at roughly 60% to 70% of the total cost as an adjustment. Buyers, however, will factor in the style and look of the flooring and be less inclined to associate an exact dollar number. So, if they fall in love with the look of the décor, they could be motivated to increase their bids substantially.

THE BOTTOM LINE IS...

If your current hardwood floor is in good shape, you aren't doing the world or your checkbook any favors by replacing it. But if you are in the market for new floors, and given the stylish, cost-effective options in eco-friendly flooring, there is no reason to consider going the old hardwood route. ■

Install
natural carpeting

ERIC SAYS Carpeting covers 70% of floors in homes across the country. It is a low-cost flooring option that hides imperfections and dirt, absorbs sound, and adds color and texture to the home. But carpeting also traps allergens, dust, and mites, and when dirt gets tracked into the carpet in your home, pesticides, bacteria, and toxic-heavy metals are brought in as well.

Carpeting is also manufactured from unsustainable and sometimes harmful materials. Synthetic carpet fibers are made from petroleum, a fossil fuel we sometimes import from hostile countries. Carpets made of nylon and polypropylene (called olefin) are usually backed in vinyl (PVC), a harmful plastic connected to a rare form of cancer. The glues, flame retardants, stain resisters, and water protectants used on carpets release VOCs into the air, which can lead to headaches, nausea, and respiratory problems. And that new-carpet smell is a toxic soup of toluene, xylene, and formaldehyde—all known carcinogens.

More than 1.8 billion sq. yd. of carpeting are sold every year—enough to cover New York City twice! But less than 1% is recycled. On average, we change carpeting in our homes every 7 to 10 years, placing nearly 4 billion lb. of carpet into landfills every year.

But manufacturers are changing their ways, offering healthier, recycled, and recyclable carpets, as well as ones made from natural materials to help lessen the impact carpet has on the planet.

Green $pecs

Overall Rating

Difficulty

Green Benefits

What will this project do for your home?

According to the EPA, the typical carpet contains at least 120 toxic chemicals, many known carcinogens or neurotoxins, which continue to off-gas from a carpet for up to three years after installation. Installing a healthy and natural carpet eliminates this exposure, as well as the runny nose, itchy eyes, and skin rashes associated with carpet fumes.

The Carpet and Rug Institute certifies eco-friendly carpets through their Green Label Plus program, which evaluates products based on VOC emissions.

What will this project do for the Earth?

The 2 million tons of carpeting dumped into the solid-waste stream account for 2% of all solid waste in U.S. landfills each year. Carpeting leaches hazardous chemicals into groundwater and soil for decades. By avoiding synthetic carpet and switching to a natural product, you'll help reduce our dependence on oil, save energy, and eliminate one of the largest sources of construction waste.

Will you need a contractor?

Installing wall-to-wall carpet can really only be done by a professional installer—carpet seams must be sewn or melded together. Check with a local carpet showroom about qualified installers.

What are the best sources for materials?

With so many types and makers of carpet, it can be hard to find a healthy one. Look for the Carpet and Rug Institute's (www.carpet-rug.org) Green Label Plus certification, which indicates carpets that are low-emitting and low-VOC.

Natural carpets are available in several materials. Wool is an ideal green carpet material—soft and long-lasting with fibers that are antiallergenic and naturally stain resistant. The dense fibers also help prevent dust from being released into the air. The strands don't break off and float like dust either, so they cannot be inhaled into the lungs.

Cotton is an alternative to wool. Hemp, like cotton, is a natural fabric fiber but requires less water and pesticides to grow. Sisal is another natural carpet fiber. Made from the agave plant, sisal is harvested without the use of pesticides or fertilizers. Its rough fibers make it ideal for high-traffic areas. Like wool, sisal is fire resistant and sound absorptive. But sisal absorbs water, making it unsuitable for damp areas.

Jute is another grass fiber. Soft and durable, it's naturally mold and mildew resistant but will biodegrade in dampness and sunlight. Jute is ideal as a natural backing for other carpet materials. Sea grass is a woven plant fiber with a distinctive grassy odor. It is highly stain resistant but also dye resistant, so it only comes in natural colors.

Linen carpeting is spun from the fibers of a flax plant to create a durable, supple, and somewhat shiny surface. Coir is an unusual carpet material made from the fibrous husk of a coconut. It is naturally insect resistant and incredibly rough and durable. It is water resistant, so it is perfect for outdoor or damp areas.

Look for carpeting with the following eco-friendly characteristics:

- Uses either a natural fiber or recycled content material for carpet strands.
- Uses natural latex rubber or jute backing instead of synthetic styrene butadiene (SB) latex, a known carcinogen.
- Uses no vinyl or PVC in the backing.
- Avoids glues like 4-phenylcyclohexene (4-PCH), which is high in VOCs.
- Manufacturer offers a recycling program to accept discarded carpeting.

If installing a carpet pad to protect the carpet fibers and create a cushy surface, avoid using synthetic pads as they are filled with harmful VOCs. Look for natural padding made from jute or recycled plastic. Padding made from recycled PVC is available, but I don't recommended it.

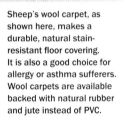

Sheep's wool carpet, as shown here, makes a durable, natural stain-resistant floor covering. It is also a good choice for allergy or asthma sufferers. Wool carpets are available backed with natural rubber and jute instead of PVC.

How much maintenance will be required after installation?

If properly cared for, a wool carpet should last 50 years. Other natural-fiber carpeting will last well beyond the typical 10 years of synthetic carpets.

How long will the project take to accomplish?

Installing carpeting can be done in a few days. Professional installers can work around your furniture, although it does slow the process a bit.

KEVIN SAYS Carpet certainly feels cozy underfoot but even so-called eco-friendly options made from recycled materials will end up in landfills sooner or later. Then there is the dust collection and regular cleaning required, not to mention the wear and tear. Often carpet is chosen to save on the cost of wood flooring, but quality green carpeting options run close to bamboo flooring and likely won't last as long.

What is the capital cost? $$$

The price of carpeting ranges vastly from $10 per sq. yd. for cheap polypropylene to as much as $90 per sq. yd. for fine wool, with plenty of eco-friendly options in the mix. That's a broad range of about $1,350 to $12,150 for 135 sq. yd. Then, of course, you add installation and padding. Installation can be a single price of $150, with custom installers charging about $5 per yd. Standard padding runs about $5 per sq. yd. (although it doesn't make much sense to use toxic padding under healthy carpet), while the eco-friendly stuff goes for $9 per sq. yd.

What financial resources are available? $

No help from the government, but there are always plenty of deals on carpet and installation. Check for specials online and be brave about negotiating. To really save money when carpeting smaller areas, ask about remnants, which can save you more than 50% of the original price.

What is the monthly cost or savings? $

Replacing a carpet every five years due to wear and tear could average out to $20 to $100 per month. Steam cleaning carpets could also cost upwards of a few hundred dollars a year.

What is the long-term home value? $

Wall-to-wall carpeting was once a desired luxury, but the novelty has worn off. Hardwood floors are considered to be more valuable in the resale environment. Appraisers never give value to carpet—green or otherwise—over hardwood floors, and buyers often figure the existing carpet is an item that they will have to replace sooner or later.

THE BOTTOM LINE IS...

If you are married to the idea of carpet in your home, pick an eco-friendly, natural option. The additional cost is minimal and you'll help out the planet. ■

Use formaldehyde-free wood for cabinetry

ERIC SAYS The natural finish of wood brings warmth into any home, but most wood products are not solid wood at all. A thin veneer of real wood is glued over a panel of compressed sawdust, reducing weight and cost and allowing the use of small, scrap pieces of wood. Unfortunately, nearly all pressed wood made in the veneering process is held together with formaldehyde.

Formaldehyde is a highly flammable and colorless gas with a pungent odor and is mostly used as a binder in pressed-wood products, namely particleboard, plywood, cabinets, and furniture. It is a known carcinogen and research shows a connection between formaldehyde exposure and leukemia as well as cancers of the nose, throat, and brain.

No matter when your house was built, you likely have formaldehyde-impregnated wood in your home. Almost every current wood product uses formaldehyde binders, which steadily evaporate into the air for years after a product is installed. Even at air levels as low as 0.1 part per million (ppm), formaldehyde can cause eye and skin irritation, burning of the nose and throat, coughing, and nausea.

Formaldehyde has been banned in Europe and Japan, but not in the United States, where wood products are a hundred times more likely to contain formaldehyde. More than 46 billion lb. of formaldehyde are produced each year, making it one of the highest-volume chemicals produced in the U.S. The U.S. also produces 10.3 billion sq. ft. of composite wood each year, releasing about 900 tons of formaldehyde emissions. Selecting formaldehyde-free wood keeps these dangerous emissions from entering your home.

What will this project do for your home?

Formaldehyde is a sensitizer and can actually make you more sensitive to other chemicals. Studies have shown that people with more than 10 years of exposure to formaldehyde are more than four times

Green $pecs

Overall Rating

Difficulty

Green Benefits

more likely to develop amyotrophic lateral sclerosis (ALS, also known as Lou Gehrig's disease) over those with no exposure. The California Air Resources Board has reported that 1 in 10,000 Californians will develop cancer from exposure to formaldehyde.

Although every composite wood manufacturer could use healthier alternatives, most choose not to. Unless more customers request formaldehyde-free products, formaldehyde will continue to be the standard. When buying any new wood cabinets or furniture, demand formaldehyde-free products to reduce the grave risks to you and your family.

What will this project do for the Earth?

Once released into the air, formaldehyde breaks down quickly into formic acid and carbon monoxide (contributing to global warming). If released into water, it dissolves quickly and breaks down. Fortunately, formaldehyde does not persist or build up in the natural environment.

Will you need a contractor ?

You will need a contractor to install new cabinets or hang a new door, but your contractor may not know if the product going into your home contains formaldehyde. Ask the manufacturer directly.

What are the best sources for materials?

Wood manufacturers aren't required to reveal if there is formaldehyde in their products. Typically, only green manufacturers using formaldehyde-free materials will go out of their way to mention the chemical.

It is a misconception that because formaldehyde is so widely used, it is a superior adhesive. Natural and healthy alternatives perform as well or better than formaldehyde woods. Formaldehyde is still in use because it is cheap, and it is cheap because it is so widely used.

If you decide to purchase wood products containing formaldehyde, look for "exterior-grade" products, which contain phenol-formaldehyde (PF) instead of urea-formaldehyde (UF). PF products generally emit lower amounts of formaldehyde gas than UF products.

Salvaged cabinets have had years to release their formaldehyde, making them a great alternative to buying new. If you're having custom cabinets built, consider using reclaimed wood. An old barn or demolished wood floor could be made into new cabinets, and both are inherently formaldehyde free.

European cabinet manufacturers comply with strict emission requirements that prohibit the use of formaldehyde. Purchasing cabinets from IKEA or directly from Europe may automatically provide you with low- or zero-formaldehyde wood.

How much maintenance will be required after installation?

Formaldehyde-free cabinetry requires the same amount of maintenance as any other type of cabinet. If using cabinets with natural oil and wax finishes, reapply the finish twice a year.

How long will the project take to accomplish?

Choosing formaldehyde-free boards may add time to a cabinet order. Plan ahead and expect an extra few weeks, just in case.

Banned in Europe and Japan, formaldehyde is still used in wood cabinets and doors in the U.S. Formaldehyde-free cabinets are a healthful alternative to those made with the known cancer-causing agent.

RESOURCES

Formaldehyde-free fiberboard:

Columbia Forest Products
www.columbiaforestproduct.com

Produces PureBond®, a green substitute for regular medium-density fiberboard (MDF) and oriented strand board (OSB).

Formaldehyde-free alternative boards for use in building cabinetry or furniture:

Flakeboard
www.flakeboard.com

Kirei
www.kireiusa.com

SierraPine
www.sierrapine.com

Cabinets built out of formaldehyde-free wood:

Breathe Easy® Cabinetry
www.breatheeasycabinetry.com

NEFF Kitchens®
www.neffkitchens.com

Uses marine-grade plywood instead of particleboard with no added formaldehyde.

Neil Kelly Cabinets
www.neilkellycabinets.com

Custom cabinets from wheat and reclaimed and recycled wood, all formaldehyde-free:

Greenline™
www.gogreenline.com

Humabuilt®
www.humabuilt.com

KEVIN SAYS We probably spend more time handling our cabinetry than touching almost anything else in our homes. It's scary to think that something so everyday could make us sick. But manufacturers have started creating healthy options for cabinets that are better for the planet as well.

What is the capital cost? $$$$

Cabinetry is figured in linear feet and can vary in cost depending upon chosen options. An average kitchen requires roughly 40 lin. ft. to 60 lin. ft., including upper and lower cabinets. Conventional cabinetry and installation can run anywhere from $250 to $600 per lin. ft. (about $10,000 to $15,000 on the low end, with a high-end kitchen running $24,000 to $60,000 for custom work). Healthy formaldehyde-free materials fall right in the middle at around $400 per lin. ft., about $16,000 to $24,000.

What financial resources are available? $

Local governments have started to use tax credits to encourage commercial builders to use formaldehyde-free materials, but these incentives haven't trickled down to the consumer level yet. Most sales on cabinetry from big-box stores are for products using formaldehyde.

What is the monthly cost or savings? $

There is no direct impact on your wallet for choosing eco-friendly cabinetry, but the health benefits may save you in the long term.

What is the long-term home value? $

The kitchen has long had a major impact on resale value for homes, and cabinetry is a major factor. At this point, there is no data to suggest that formaldehyde-free cabinetry attracts a higher price in the marketplace, but it certainly won't detract for health-conscious buyers.

THE BOTTOM LINE is...

There are plenty of eco-friendly choices for cabinetry. If there is a healthy option available at a comparable cost, then that means there's zero downside. ∎

Build a composite material deck

ERIC SAYS Decks are a welcome extension of a home, providing a connection to the outdoors and adding beauty and value to property. Unfortunately, decks also add to the ongoing environmental crisis of deforestation and can be a maintenance headache.

Most decks are built from wood harvested by logging—some 3.3 billion lin. ft. of decking are made each year. That's enough to circle the Earth more than 25 times. And although wood is a natural material and can be replanted, the rate of logging far outweighs reforesting efforts.

The resulting deforestation is staggering. An area of forest larger than the state of Pennsylvania is logged every single year. Global deforestation is responsible for approximately 20% of global warming, and more than 80% of ancient forests on the planet have been logged and destroyed. What remains is at risk to be gone by the end of the century.

Most decks are built from pine, cedar, or redwood trees. Cracking, warping, and splitting occur over time and affect the appearance and structural integrity of a deck. Decks need to be sealed annually and maintained to prevent rot, mold, and mildew.

Building a deck with a green alternative material slows deforestation and the amount of maintenance is significantly reduced. Decks of plastic lumber, recycled from the 27 million tons of plastic discarded each year, make good use of a waste product. It never cracks or warps, it doesn't require sealing, and it cannot rot. Composite wood is a mix of recycled plastic and sawdust (the largest waste product from the lumber industry). Composite boards look like real wood and don't expand and contract with temperatures as much as plastic lumber.

Using composite decking helps reduce the rate of deforestation, but it also encourages the use of plastic, a nonrenewable resource. Eventually, that recycled plastic decking may end up in a landfill. Despite these concerns, I continue to recommend composite decking because it creates a durable, long-lasting, and recyclable material.

Green $pecs

Overall Rating

Difficulty

Green Benefits

Maintaining wooden decks (above) involves sealing with chemicals that are toxic and, because they are oil based, further our dependence on fossil fuel. Compared with a traditional wood deck, composite wood (right) requires significantly less maintenance and is free of splinters, cracking, and splitting.

What will this project do for your home?

Wood decking requires stripping, staining, and sealing every year, while a composite deck only needs a good cleaning. Instead of harsh chemicals and exposure to toxins, all you'll need is a wire brush.

Composite decks are also impervious to termites and water damage. Wood decking might last 10 years, but a composite deck could last two to three times longer (they haven't been around long enough to know for sure, but every composite deck manufacturer makes this claim publicly).

What will this project do for the Earth?

The construction of a typical deck (about 25 ft. × 15 ft.) requires an entire tree's worth of wood. Using composite decking material prevents this logging and the release of the equivalent of 400 lb. of carbon emissions.

By decreasing demand for wood and exotic tropical hardwoods such as mahogany, cypress, and ipê, illegal logging practices in the Amazonian rainforest are lessened and the rate of destruction of tropical forests (now on course to last only another 35 years) is combated.

Will you need a contractor?

The complexity of pouring foundations and fastening structural connections makes building a deck complicated. Do yourself a favor and hire a professional.

What are the best sources for materials?

Composite and plastic lumber are available from dozens of companies, but be careful. Many call products green simply because they don't contain wood. Many use new (called "virgin") plastics or PVC plastic. PVC (vinyl or polyvinyl chloride) is an energy-intensive plastic that is hazardous throughout its entire life cycle, from production through disposal. Avoid plastic lumber containing PVC, even if it's recycled PVC. Instead, look for products that use 100% recycled plastics.

Both plastic and composite lumber do not rot, but when in direct sunlight, plastic lumber can get hot enough to burn bare feet. Composite lumber will be cooler to the touch, but its color will fade over time. Otherwise, the two are similar in installation and maintenance. The higher the plastic content of the decking, the more it will expand and contract with temperature changes.

Hidden fasteners, perfectly sized planks, and easy installation make composite wood an ideal choice for a low-maintenance deck.

Some decking planks are solid boards to add strength; others are hollow to reduce weight and material. Plastic and composite decking come in several shapes and profiles and can be bent into curves to create unusual designs.

Other installation options, such as blind or hidden fasteners, are available, which eliminate the need for exposed screw heads on the top of the deck. Hidden fasteners also make the deck easier to clean and avoid staining from rusting screws.

Although the decking surface can be made from composite lumber, the supporting structure below must still be made from pressure-treated structural wood members. Pressure-treating is a method of preserving wood for outdoor use. Pressure-treated wood was originally treated with chromated copper arsenate (CCA), which has since been banned by the EPA. Some CCA-treated wood is still sold for commercial use, so be careful. CCA wood releases arsenic into the skin when touched. Studies have shown treated wood as old as 15 years continues to release high levels of arsenic into the air and surrounding soil. (If you're worried your existing deck contains wood treated with CCA, order a CCA test kit like the one available from Safe2Play [www.safe2play.org].) For any new deck structure, try to find wood treated with the healthier alkaline copper quaternary (ACQ). Wood that has been pressure-treated with ACQ must use stainless or galvanized screws and nails.

How much maintenance will be required after installation?

Despite claims of many manufacturers, plastic and composite decking are not completely maintenance free. Although it never needs stripping, staining, or sealing, composite decking should be cleaned every year with a wire brush or power washer. Chemical cleaners may leave a white residue on your deck, so avoid them. The manufacturer will be able to provide instructions on proper product care and maintenance.

How long will the project take to accomplish?

Plastic or composite decking does not take any more time to install than traditional wood decking. In fact, hidden fasteners may speed up the process and, unlike real wood, composite decking is consistently straight and perfect without knots, crooks, or other defects common to regular wood.

KEVIN SAYS The surest way to engage in a heated argument is to bring up the topic of composite decks. Many in the green community swear by them since they use recycled plastic and help protect virgin forests from being cut down for mere weekend leisure. But contractors and consumers have raised havoc on the Internet and in the courtrooms with complaints and lawsuits over the materials not being as perfect as represented. Many complain that materials were not sufficiently tested before being brought to market and that the natural materials (like sawdust) used in the manufacturing process end up deteriorating despite the mixed compounds. This can result in mold, flaking, discoloration, and, in some cases, disintegration in as little as two years. There have been a number of settlements in these suits, so buyers beware.

What is the capital cost? $$

We priced a rectangular deck of 16 ft. × 20 ft. Since much of the decking material comes in 16-ft. lengths, it makes for little waste and an easy comparison. Assuming labor, permits, sealant, railings, and the support structure being equal at around $2,800, we'll leave the major comparison to standard pressure-treated materials versus composite materials. Using 6-in. widths, the deck requires 40 boards, or 640 lin. ft. Big-box prices on standard pressure-treated lumber run about $1.30 per lin. ft., bringing the total cost to $3,632. Most of the price lists showed composite materials at a similar price of $2 per lin. ft. This would put an eco-friendly deck price at $4,080, or a premium of less than 15%.

What financial resources are available? $

We were unable to find any government support for deck building, eco-friendly or otherwise. The materials are not generally subject to big sales, although deck companies have specials from time to time.

What is the monthly cost or savings? $

Wood decks require maintenance on an annual basis. A couple of days of work and $150 in materials each year will be all that is necessary to keep the deck looking good for about 15 years. There is debate as to whether composites require similar treatment. Some composite companies suggest simple maintenance, while most claim not to require or allow for sealants. While many claim the decks last longer than 20 years, it's too early to tell. Most complaints about composite materials stem from the decomposition of wood products mixed with the plastic. Untreated, the natural materials used in composite wood can mold, splinter, and discolor—degenerating the deck. All in all, there is not enough data to support the claim that composite decks are maintenance money savers.

What is the long-term home value? $$$

Attached decks are an affordable way to add space to your home in areas where outdoor living is an attractive option. Generally, appraisers will give a deck an additional value of 60% to 70% of the cost, depending upon function and condition. Buyers will respond positively to a well-designed deck that compensates for lack of family space and may pay more when comparing like models. There is no data to support that buyers will pay more for an eco-friendly option, and they may in fact become concerned about the use of composite materials if they read about the controversy with these materials on the Internet.

THE BOTTOM LINE IS...

Having a deck is a great way to enjoy the outdoors and add long-term value to your home. Stay tuned about the debate on composite deck materials, as consensus has still not been reached. If it turns out that composite decking consistently stands the test of time, then you can feel good about all that plastic bottle recycling. Until then, choose your composite carefully. ▪

RESOURCES

Recycled plastic decking:

Canopy Plastics
www.canopyplastic.com

Cascades Re-Plast, Inc.
www.cascadesreplast.com

Everlast Plastic Lumber
www.everlastlumber.com

PlasTEAK® Inc.
www.plasteak.com

Plastival Inc.
www.plastival.com

Renew Plastics
www.renewplastics.com

Rumber®
www.rumber.com

Trelleborg
www.trelleborg.com/
ecoboard

Composite lumber decking:

ChoiceDek®
www.choicedek.com

Master Mark Plastics
www.rhinodeck.com

Nexwood
www.nexwoodnw.com

TimberTech®
www.timbertech.com

Trex
www.trex.com

Harvest your rainwater

Green $pecs

Overall Rating

Difficulty

Green Benefits

ERIC SAYS Water is one of our most valuable, yet most endangered, natural resources. Only one-third of water used in a home needs to be drinkable, but we use potable water for every single drop consumed in and around our homes. While we need to have clean water to drink, cook, and bathe, our other daily activities don't require fresh water. We could use rainwater instead. The idea of harvesting rainwater has been around for thousands of years. Modern technology has just made the process easier and safer.

Rainwater harvesting is the process of collecting the water that falls on the roof or yard. The water is filtered and stored for later use. A catchment system can be used to collect water to irrigate a garden, water a yard, wash clothes, or flush toilets. With a purification system, the water could even be used for drinking (although this is rare). Rainwater harvesting (or catchment) ranges from simple rain barrels in the yard to complex systems that pump, filter, and store water.

What will this project do for your home?

If you pay for city water, harvesting rainwater lowers water bills and provides better water for plants. Even unfiltered, rainwater is superior to groundwater. It is considered "soft" water, ideal for irrigation and watering plants. If you use well water, harvesting reduces the demand on the local watershed and helps with low-flowing wells.

What will this project do for the Earth?

Somewhere in the world, someone dies every eight seconds due to thirst or waterborne disease. More than 1.1 billion people do not have access to safe drinking water. By catching rainwater, you reduce the strain on the global water supply and on your local public water system. You also reduce the demand on public water treatment plants and on local sewer and storm water collection systems.

Will you need a contractor?

For a basic garden water collection system, you can simply purchase rain barrels and install them under downspouts around your home. Such systems can be installed by one person in less than an hour, and when it rains the water is fed straight into the barrel. A more extensive system will require the help of a professional.

What are the best sources for materials?

There are five components of a rainwater-harvesting system:

1 **Surface:** Typically, rainwater is collected from the surface of a roof, but driveways and paved areas can also be used. Asphalt roofing shingles poison rainwater, leaching oil into the water and making it unusable for gardening or laundry use. Roofing materials such as tile, metal, and plastic do not absorb or taint water.

2 **Collection:** As it rains, water is collected in a roof's gutter, falls into the downspout, and is dumped on the ground. The mouth of a downspout is the easiest place to install a storage tank (rain barrel or cistern). Water collection can also be done through a roof drain and fixed pipe. Nearly any method is fine as long as the water ends up in a storage tank.

3 **Storage:** The heart of a rainwater-harvesting system is the storage tank. Tanks come in a number of sizes and materials, but remember that water weighs more than 8 lb. per gal., so you'll want to locate the tank in a place that can support its weight when full. Size the tank to accommodate a reasonable period of rainfall. If you want to use the water for potable use, you must purchase a tank with a United States Department of Agriculture (USDA)–approved lining marked "food grade." Be careful when purchasing recycled or used storage tanks. They may contain lead or other harmful chemicals. Make sure they were used to store only water and nothing else. Tanks range in size from 55-gal. barrels up to 15,000-gal. containers.

   In freezing areas, large tanks must be kept indoors or below ground. Smaller rain barrels can simply be left open in the winter to prevent freezing. Dark tanks absorb heat and prevent algae growth inside the tank. Fiberglass tanks are durable, and leaks are easily repaired. Tanks made of polyethylene terephthalate (PET plastic) are the most common but have a tendency to leak. Both types can be placed above ground or below ground, and both must be lined.

4 **Filtering (optional):** Rainwater harvested from roofs may contain animal or bird feces, dust, smog, or pesticides. As it washes down the surface of the roof, the water picks up additional dirt and large particles. That's why all rainwater should be filtered before being used. To be suitable for drinking, rainwater must go through a series of filters and an ultraviolet light that kills virtually all bacteria.

The first rush of rainwater off a roof will contain high amounts of pollutants; many filters have a mechanism to flush out the first few gallons of collected water. Called a "first-flush" diverter, this valve dumps the water to keep pollutants from going through and overloading the filters.

5 **Distribution (optional):** Stored and filtered water needs to be sent to where it will be used (laundry, garden, toilet). A simple rain barrel can have a small valve and hose attached to water your garden. A pump may be used to distribute the water around the house. More extensive systems, however, require a network of pipes to send the water to the appropriate locations. Extensive distribution systems are difficult to install in existing homes, unless you're doing a complete remodel.

Any stored water should be kept covered to prevent the growth of algae and mosquito breeding. All tanks should also use mosquito dunks, which are small discs that release a bacterium called BTI (*Bacillus thuringiensis israelensis*) into the water. BTI is harmless to people, fish, birds, and animals but kills mosquito larvae. Dunks are available at gardening and home center stores and online at Amazon.com.

At least 36 U.S. states expect to face a water shortage in the next decade, regardless of how much it rains. With a rainwater-harvesting system (like Rainwater HOG, www.rainwater hog.com), water that falls on the roof is directed into a storage tank and conserved for eventual use in and around the home.

How much maintenance will be required after installation?

Water-catchment systems should be checked frequently for tank leaks. Gutter screens should be cleaned periodically, and storage tanks should be drained and cleaned annually.

How long will the project take to accomplish?

Installation of an extensive rainwater-harvesting system will vary based on the location and size of the tank. A typical system will take about a week to install in a new home.

KEVIN SAYS If you live in the West and install a rain barrel or water-catching system, you might be considered a common criminal. There are heated debates in Colorado, Utah, and Washington about the legality of harvesting rainwater. It seems that rainwater in these areas is considered property of the state as groundwater that ultimately belongs to the streams and rivers feeding agriculture. Although the tides appear to be turning in favor of conservation, keep in mind that retrieving rainwater on a large scale could create changes in water flow that affect already drought-impacted areas. Not that the water police will be visiting granny's garden anytime soon.

The big question: Is it worth the time and effort? Those with asphalt shingles will have to reroof at a big cost just to consider rainwater harvesting. Then there are the issues around sporadic rains. If it rains a lot in your area during a particular season, you won't benefit much since the rain itself waters the plants, and when it's dry you won't have much water in storage to use. Other complaints about barrel systems center on challenges of low pressure or having to carry water to irrigate. More advanced systems use pumps, but they use energy and add cost.

What is the capital cost? $$$

Rain barrels are not all that cheap. Prices for a 50-gal. barrel range anywhere from $50 to $200, with the entire system costing anywhere from $2 to $5 per gal. of storage depending upon its sophistication and installation needs. A 1,000-gal. system could cost anywhere from

Estimating Rainwater Collection

The average house is about 2,000 sq. ft. With two floors, each floor is about 1,000 sq. ft. As a general rule, 1,000 sq. ft. of roof collects 632 gal. for every inch of rainfall. This is true anywhere in the world.

For example, Chicago receives 35.82 in. of rain a year. Multiply that by 632 gal. per inch of rainfall, and you get 22,638.24 gal. So a 1,000-sq.-ft. roof in Chicago can collect 22,638 gal. of rainwater a year. A 15,000- to 20,000-gal. tank would be the largest you should consider in this case.

The chart below indicates potential rainwater harvesting on a 1,000-sq.-ft. roof based on the annual rainfall of the cities listed.

CITY	ANNUAL RAINFALL (IN.)	HARVESTED WATER FROM 1,000-SQ.-FT. ROOF (GAL.)
Mobile, AL	63.96	40,423
Juneau, AK	54.31	34,324
Phoenix, AZ	7.66	4,841
Little Rock, AR	50.86	32,144
Los Angeles, CA	12.01	7,590
San Francisco, CA	19.7	12,450
Denver, CO	15.4	9,733
Washington, DC	38.63	24,414
Miami, FL	55.91	35,335
Atlanta, GA	50.77	32,087
Honolulu, HI	22.02	13,917
Boise, ID	12.11	7,654
Chicago, IL	35.82	22,638
New Orleans, LA	61.88	39,108
Boston, MA	41.51	26,234
Detroit, MI	32.62	20,616
Reno, NV	7.53	4,759
Albuquerque, NM	8.88	5,612
New York, NY	47.25	29,862
Charlotte, NC	43.09	27,233
Portland, OR	36.3	22,942
Philadelphia, PA	41.41	26,171
Dallas, TX	33.7	21,298
Seattle, WA	37.19	23,504

(Adapted from data from 2000 U.S. Census—30-year data average from 1971–2000)

$2,000 to $5,000. Installing pumps and fittings to make life easy could add another $2,000.

What financial resources are available? $$$

Many local governments and utilities have some sort of rain-collection incentives. Cities such as Tampa, Florida, simply provide free barrels, but states such as Arizona provide tax credits for 25% of cost up to $1,000. Contact your local water district office for information.

What is the monthly cost or savings? $

Water savings vary widely depending upon area rainfall. Best estimates show that with a complete system, a 30% savings is optimistic. The national average cost for city water is a penny and a half per gal. If using 2,100 gal. of water a month, the bill runs about $31.50. A 30% savings would only give you an extra $9.45 in your pocket. That means it would take you 17.6 years to recoup the cost of a low-end system. You may have additional maintenance to keep the system clean and running, extending the time it takes to recoup the investment.

What is the long-term home value? $

Although catchment systems have been around forever, they haven't been a part of the mainstream housing market. No appraisers we found would give additional value to a home for having a water-reclamation system. Since most buyers are unfamiliar with the systems, it's unlikely they will pay more for a home that has one, although there is no readily available data to support either side of the argument. Concerns about additional maintenance can sometimes be a detractor, so the responsibility of educating the buyer will be incumbent on the seller.

THE BOTTOM LINE is...

If you have a roof made of something besides asphalt shingles, live in a high-rain area, and have the space for the tanks, you could consider a water-catching system to help the planet along and feel good. Just don't expect to be able to brag about all the money you are saving from a rainy day. ▮

RESOURCES

Rain barrels:

**Green Culture
Wonder Water**
www.wonderwater.net

Oak Barrel Winecraft
www.oakbarrel.com

Rain Barrel Source
www.rainbarrelsource.com

Real Goods
www.realgoods.com

Rainwater tanks:

Aquadra Systems
www.aquadrasystems.com

Bushman USA
www.bushmanusa.com

The Garden Watersaver
www.gardenwatersaver.com

Norwesco, Inc.
www.norwesco.com

Rain Harvesting Pty, Ltd.
www.rainharvesting.com

Rainwater HOG
www.rainwaterhog.com

Snyder Industries
www.snydernet.com

Water Filtration Company
www.waterfiltrationcompany.com

Disassemble, don't demolish

Green $pecs

Overall Rating

Difficulty

Green Benefits

ERIC SAYS Every year in the U.S., approximately 1.75 billion sq. ft. of buildings are torn down and literally thrown away. That's 1,900 houses pulverized into rubble and dumped into a landfill each day. In addition, the Environmental Protection Agency estimates that 164 million tons of new construction waste is produced every year in the U.S., equivalent, in materials, to 250,000 homes thrown away annually. It accounts for 35% to 40% of trash in landfills, only 20% of which is recycled.

Instead of demolishing old buildings, however, we can disassemble and salvage valuable materials from them, saving wood, steel, glass, fixtures, and appliances that can be sold or reused in new buildings. And through construction-salvage techniques, almost a billion board feet of lumber could be saved and reused each year.

What will this project do for your home?

A typical demolition crew is similar to a pack of piranhas—they go into a building and destroy everything marked to go. They are fast, messy, and produce a pile of debris you'll need to pay to haul away. Deconstruction (or disassembly or salvage) is a more careful, systematized approach to disassembling a building in order to extract reusable items and materials.

What will this project do for the Earth?

Building a house produces from 4 lb. to 8 lb. of waste per sq. ft. By recycling 85% to 95% of the wood, metal, glass, and cardboard from a construction project, at least 4 tons of waste is kept out of the nation's crowded landfills. The reduction in energy and carbon emissions from recycling just one home's construction waste is equivalent to taking one car off the road for a year.

Construction Waste Generated from a 2,000-Sq.-Ft. Home

Building a new home produces 8,000 lb. of waste, enough to fill a bedroom from floor to ceiling with materials. More than 85% of this material could potentially be recycled. This chart indicates the weight and volume of the typical waste generated from building a 2,000-sq.-ft. home.

MATERIAL	WEIGHT OF WASTE (LB.)	VOLUME OF WASTE (CU. YD.)
Solid sawn wood	1,600	6
Engineered wood	1,400	5
Drywall	2,000	6
Cardboard	600	20
Metal	150	1
Vinyl/PVC (from siding)	150	1
Masonry (bricks)	1,000	1
Hazardous materials	50	—
Other	1,050	11
Total waste	8,000	51

(Data adapted from the National Association of Homebuilders)

Will you need a contractor to complete this project?

Unlike demolition, which requires a crew with heavy equipment, deconstruction can be done by anyone looking to extract materials for reuse. Only finishes, fixtures, and equipment should be removed. Leave extracting anything structural to the professionals.

If you are working with a contractor, request that he or she salvage and recycle construction waste. Local and state waste management boards can provide information and resources on how and where to recycle the materials. Ask the contractor to save the receipts and documentation of the recycling. You'll need them if applying for a green building certification for your home, such as Leadership in Energy and Environmental Design (LEED; www.usgbc.org).

What are the best sources for materials?

Construction disassembly has four stages. The first stage is salvaging and extracting easily removed items such as doors, windows and skylights, light fixtures (recessed and hanging), appliances, cabinets, sinks, faucets, bathtubs, moldings and fireplace mantels, ceiling fans, railings, hot water heaters, and wood flooring. Make sure to note any materials or finishes that need special care when being removed. Mark anything you want to keep and reuse yourself, so the contractor does not take it

DO THE MATH

Remodeling or building a home produces a great deal of waste. More than 85% of this material could be salvaged, reused, or recycled.

by mistake. Salvaged items can then be reused or picked up by a local chapter of The Reuse People (www.thereusepeople.org) or Habitat for Humanity (www.habitat.org). You might also find a local salvage yard to remove and haul away items for free.

Cleared of fixtures, the remaining building can be taken apart. This second stage is called deconstruction. Instead of using a bulldozer, wrecking ball, or sledgehammer, hand tools are used to take apart a building piece by piece. This keeps usable materials intact and easier to recycle.

The third stage involves sorting materials so they can be hauled away. (Sorting should be done at both deconstruction and new construction sites.) As debris is pulled from a building, it should be sorted into piles. This makes it easier to find scraps that could be needed during the construction process, and when construction is done, sorted materials are much easier (and twice as likely) to be recycled. Your contractor may have his or her own methods of doing this, but I suggest you make separate bins for scrap wood; drywall; cardboard; metal, glass, and plastics; vinyl/PVC siding (recyclable, but few places accept it, so call your local waste management board to find out about options); masonry (old masonry can be reused in a garden or as backfill against a new foundation), stone, and brickwork; hazardous materials (asbestos and lead paint should be handled by professionals, and fluorescent lights should be disposed of properly because they contain mercury); unrecyclable construction waste; and compost materials. You can also

contact your local recycling center for more information on how materials should be sorted.

The last stage of disassembly is the actual recycling. Take the sorted bins to a nearby recycling center instead of a landfill. Deconstruction does not have to be a painstakingly slow process; it just involves a little more careful and thorough planning during demolition.

How much maintenance will be required after installation?

If you have an architect or contractor, he or she will be able to help plan your deconstruction. Deconstruction may require the temporary storage of select materials, fixtures, and appliances.

How long will the project take to accomplish?

Construction salvage and recycling requires extra effort and time. Plan accordingly in your construction schedule for deconstruction to take about one and a half times as long as traditional demolition. Instead of a single receptacle for waste, you'll need several collection bins to remain in place for the duration of the process.

KEVIN SAYS My first ever mortgage client was a demolition contractor who took apart old houses in Los Angeles. He built a fabulous home completely from materials he had taken from his deconstruction job sites. It's about time the rest of the world caught up with his approach. It's a shame to waste perfectly good material when there is an easy alternative. More and more homeowners are catching on, planning deconstructions to save on building costs and offset demolition costs.

What is the capital cost? $$$

As deconstruction becomes more popular and more companies enter the business, prices have come down. What you pay for in this process is the labor. While deconstruction labor may be two to three times that of demolition, it can be offset by a number of factors. No heavy equipment rentals or skilled labor for operating heavy machinery is required. Additionally, there are fewer transport and dump fees, and some

RESOURCES

Habitat for Humanity
www.habitat.org

The Reuse People
www.thereusepeople.org

salvagers will pick up materials as well. For a remodel or teardown, careful deconstruction can also save landscaping. All in all, a house that requires $5,000 for demolition may only cost a net of $10,000 for deconstruction after salvage.

What financial resources are available? $$$$

The largest benefit of deconstruction comes from donating materials to 501c3 nonprofit charities. A donation of materials valued at $50,000 could qualify you for as much as $12,000 in federal tax savings. The savings depend upon income and may need to be taken over several years. Still, savings are savings. State deductions may also be available. Start by checking in with a certified public accountant when considering this approach.

What is the monthly cost or savings?

There is no monthly cost or saving to be considered in the deconstruction process.

What is the long-term home value? $

While long-term value may not be impacted directly through deconstruction, saving money by reusing materials helps provide a lower cost to recoup for a remodel.

THE BOTTOM LINE is...

Deconstruction is a smart idea. Financially, it makes sense to reuse materials, save the environment, and—who knows, with the right donation and accountant— you might even come out ahead on the deal. ∎

13 green home projects you can do when building new

Building a new home provides greater opportunities for greening than can be achieved when not starting from scratch. The projects are integrated into the structure and central systems of the home to help use water, energy, and natural resources in a new and efficient manner.

Reclaim your water, gray and hot

Green $pecs

Overall Rating

Difficulty

Green Benefits

ERIC SAYS Nearly every single drop of water that was present when Earth formed billions of years ago still remains today. The water in your glass may have been consumed by Thomas Jefferson or dropped as rain over the Egyptian pyramids. Water is not created—it is in constant motion, changing from liquid to gas to solid, and back and forth again, over and over and over.

Earth has reused all of its water for billions of years, cycling, filtering, and cleaning it naturally, but the industrial activities of humanity have interrupted the water cycle. Only 1% of water on the planet is available for us to drink (the rest is either frozen or seawater). Nearly 70% of available water is chemically polluted. Although we are not running out of water, we are running out of clean drinking water. We must learn to reuse water the way nature does.

What if we could use the water in our homes more than once? A home is a large consumer of water, but only half of it needs to be drinkable. Showers, baths, faucets, and laundry require clean water, but the soapy water going down the drain is called gray water. Although not safe for drinking, gray water is safe to use in toilets and to irrigate. The use of gray water dates back to rancher days, when farmers saved bathwater for irrigation. But, due to health concerns and antiquated building codes, gray water use has been limited in modern times.

Named for its grayish cast, gray water accounts for around 60% of water flowing out of a home and it can be reused in areas that don't require potable water—up to half of household water is used in the yard and could be replaced with gray water. Another 10% to 15% is flushed down the toilet. Since no one is drinking out of the toilet, gray water could be used instead. A gray water system collects wastewater from select drains around the house and stores it for a second use; it could cut the waste of drinking water by up to 60%.

Whole-House Gray Water System

A gray water system collects the soapy water from bathtubs and laundry machines and reuses it in gardens or toilets.

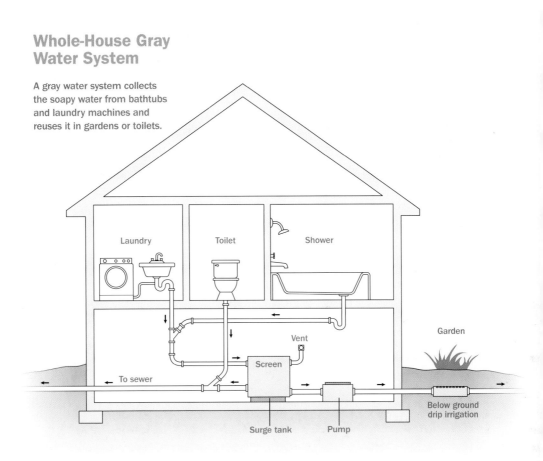

Most gray water is first heated for use in a shower or laundry. That heat can also be recovered and reused, just like the water. A drain-water heat-recovery system (DHR) is a special drain installed in a tub or shower that steals the heat from gray water and recirculates it back to heat the water in the shower. The addition of a DHR can cut hot water energy use by 20%.

What will this project do for your home?

A gray water system cuts water usage and can also lessen the strain on a septic system. Since gray water contains small traces of nutrients and bacteria, it is ideal for irrigation and fertilization of lawns, plants, and gardens.

The nation's wastewater plants and drinking water systems spend about $4 billion per year on energy to treat water.

A 10% reduction in U.S. drinking water and wastewater systems, realized through cost-effective investments, would collectively save approximately $400 million and 5 billion kWh annually.

**If 10% savings of water = 5 billion kWh and $0.4 billion,
then 60% savings of water = 30 billion kWh and $2.4 billion.**

See http://www.energystar.gov/index.cfm?c=water.wastewater_drinking_water for more info.

Almost 90% of hot water goes down the drain, but that heat can be recovered. A drain heat-recovery system can recapture most of this heat and lessen the energy used by your hot water heater.

What will this project do for the Earth?

Scientists estimate that by 2035, roughly two-thirds of the world's population will not have access to safe drinking water. Each day, 55 billion gal. of fresh drinking water are flushed down the toilet, while billions of people go thirsty.

The processing of water (filtration, treatment, pumping) requires 50 billion kilowatts of energy each year in the U.S. If every home utilized a gray water system to reuse water, it could cut this energy use in half and prevent the equivalent greenhouse emissions of nearly 4 million cars.

Will you need a contractor?

The design of a gray water system requires an experienced engineer and plumbing contractor. Online, you'll find groups of gray water hobbyists and do-it-yourselfers, but be careful of the advice you get. Every system and installation has its own unique requirements that may not fit the experience of other enthusiasts. The actual plumbing installation itself is fairly easy for someone with plumbing experience. Otherwise, leave it to the pros.

A DHR system installs easily in place of a regular shower drain. The fixture can be purchased commercially and installed by any plumber. Although DHR systems are not in widespread use, a good plumber will be able to follow a manufacturer's instructions.

What are the best sources for materials?

Before you get too excited about installing a gray water system, know that they are not permitted by many building departments. Most are unfamiliar with gray water systems and prohibit their installation due to unfounded health concerns. In reality, a properly designed gray water system poses no health risks. The water is filtered and never comes into contact with humans. But the use of gray water in toilets is a particular concern to most building inspectors. Your first step is to call your local building department and ask if they approve the use of gray water in residential toilets. If not, they may still allow its use for irrigation, so long as you use a drip system (see p. 70). Never spray or use a sprinkler with gray water. If collecting gray water from a laundry, avoid using detergents containing phosphorous, as they harm plants and pollute the watershed.

Gray water systems generally consist of a diverter valve, treatment tank with a natural sand filter, and storage tank. If gravity does not provide enough pressure, an electric pump is used to move the water through the system. From the storage tank, the cleaned water is sent to the toilets or irrigation system.

Unfiltered gray water contains trace amounts of hair, skin cells, and dirt (from showering), but they are removed through a filter. To satisfy the local building department, a bypass valve is usually added to allow you to direct gray water to the sewer. If you're still concerned about health risks, an ultraviolet light can be installed that kills virtually 99.9% of all bacteria. But no matter how clean it gets, gray water is not intended for drinking or bathing.

Typically, a gray water system gathers wastewater from bathtubs, showers, washing machines, and bathroom sinks. Kitchen sinks are not recommended since food waste is commonly put down the drain. Retrofitting an existing home with a full gray water or DHR system is difficult. If you have a one-story house with a full basement or crawlspace, you may be able to easily access existing pipes and have a simple installation. Otherwise, tearing up slabs and walls is a lot of effort for installing a gray water system.

For existing homes, you can add individual gray water systems instead. The AQUS system (www.watersavertech.com) consists of a small box that sits under a bathroom sink and pumps sanitized water, collected from the sink faucet, to the adjacent toilet (see p. 18). Envirosink® (www.envirosink.com) is a small accessory sink that attaches to an existing kitchen sink. As you wash your hands over the Envirosink, the soapy water is stored in a small tank underneath the sink. The water can be used in your garden.

A DHR looks like a standard shower drain but is wrapped in copper coil (see the drawing on p. 212). As hot water goes down the drain, the heat is transferred to clean water in the copper coil. From there, newly heated water is circulated to the showerhead or hot water heater. The DHR extends the amount of hot water available and greatly reduces the energy used to heat it in a hot water heater. A DHR works with all types of water heaters, including on-demand and solar water-heater systems.

Drainwater Heat-Recovery System

A drainwater heat-recovery system recycles the heat from soapy shower water as it goes down the drain. The heat is used to warm cold water in the coils, which is then recirculated back to the showerhead.

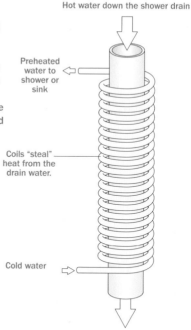

Hot water down the shower drain

Preheated water to shower or sink

Coils "steal" heat from the drain water.

Cold water

Cooled water continues down the drain.

How much maintenance will be required after installation?

A gray water system requires minimal monthly maintenance. Each month, the filter needs to be cleaned and a large chlorine tablet (resembling a giant aspirin) needs to be dropped in. The tank should be drained and cleaned annually to prevent salt buildup. The manufacturer will be able to explain exact maintenance requirements.

If left unmaintained, a filter could clog and slow the flow of water. The system will automatically use fresh water instead, negating savings.

The DHR system is maintenance free and has no moving parts. Once installed, you'll never need to worry about it. It should last 50 years.

How long will the project take to accomplish?

Whole-house gray water systems must be installed at the same time other plumbing goes in a home. Instead of every drain going to the sewer, a handful of drain pipes will be diverted to a storage tank. This does not add any significant time to overall construction. Coordinate the delivery of equipment with the contractor to avoid any delays. Single gray water systems take two to three hours to install and can be done after construction has finished.

DHR systems must be installed at the same time as the shower. Each drain requires several hours to install, but this will not slow the construction process. Other trades and subcontractors will be able to continue work during the installation.

KEVIN SAYS The concerns about gray water systems stem from the bacteria, salts, and chemicals that may come from the wastewater and seep beyond your property lines. However, the issues surrounding health concerns about gray water usage have given way to public approval. Arizona, New Mexico, and Texas have all passed laws approving gray water systems, and California is heading down the same path. Many plumbers are anticipating new business opportunities from installing gray water systems, and the plumbing trade is reexamining the Uniform Plumbing Code to potentially incorporate gray water systems. As long as the water moves straight to the garden or toilet, apparently most people don't care where it's from.

What is the capital cost? $$$

A gray water system is relatively affordable. The system itself costs roughly $1,000 to $2,000 plus installation, which takes about a day. The installation cost runs as much as $2,500, depending upon the local water-treatment system. Total system costs run from $3,500 to $4,500. A DHR system may add another $350 to $500 per drain plus another $300 to $500 for installation for a total of $650 to $1,000.

What financial resources are available?

Since Arizona is the first state to officially legalize gray water systems, it makes sense for the state to promote it financially. Arizona offers a tax credit of 25% of the cost of the system, up to $1,000. Check with your local municipality to see what is offered for both gray water and DHR systems.

What is the monthly cost or savings? $$$

Gray water from sinks and showers probably won't be enough to
continually support all of the toilet and irrigation needs in a home.
Depending upon usage, however, you could see as much as a 30% to
60% decrease in water bills. If you spend $150 monthly, you could real-
ize a $45 to $90 savings. But a gray water maintenance contract may be
required by local municipalities for monitoring safety issues at a cost
of roughly $50 per month. The net effect is that paying for the system
could take anywhere from six to nine years.

The DHR may save 20% to 40% of a water-heating bill, or roughly
$10 to $25 per month, depending upon usage. In the best case, a DHR
system pays for itself in a little more than three years.

What is the long-term home value?

Several appraisers stated there was no downside to adding a DHR
system, but there is no data to support that buyers are paying extra for
homes with one installed. The same is true for data on gray water systems, and there is still concern among consumers about health risks,
which could have negative impact if the buyer is not convinced about
the safety of the system. As more states accept the systems, the fears
should subside.

THE BOTTOM LINE IS...

DHR systems are cheap and easy when building a new home. With a
short payback, it's worth considering. Gray water needs government
approval and social awareness and acceptance for the benefits to outweigh the uncertainties. It probably won't be worth the cost in an
existing home, but when building a new house, you can always structure the water system so that a gray water system could easily be
added later on. ∎

Substitute fly ash for cement in concrete

ERIC SAYS Concrete is one of the most common building materials on Earth. It is strong, versatile, durable, and fire resistant. It is made from natural materials and does not release any fumes, toxins, or VOCs. If that wasn't enough, concrete is even recyclable! It sounds like a green material if ever there was one, so what's the catch?

Concrete is a mixture of several materials—pulverized stone, sand, a little bit of water, and 10% to 15% Portland cement—the binder that holds concrete together and gives it strength. Unfortunately, the production of Portland cement has serious adverse effects on climate change.

Cement production is one of the most energy-intensive industries on the planet. It begins by mining raw minerals out of the ground, with nearly 1.75 tons of raw materials needed to produce just 1 ton of cement. The materials are then crushed and heated up to 3,000°F. The resulting powder is ready for mixing into concrete, stucco, or grout. It's a simple process, but the quarrying, crushing, and heating require an incredible amount of energy—producing 1 ton of Portland cement releases 1 ton of carbon dioxide. More than 110 million metric tons of Portland cement are consumed each year (releasing the same amount of CO_2 into the atmosphere). The production of cement is directly responsible for 8% to 10% of all man-made carbon emissions and is the largest single industrial source of global warming.

Two thousand years ago, ancient Romans built incredible monuments out of concrete without any Portland cement at all. Structures like the Pantheon were made with lime and volcanic fly ash to create lightweight concrete that is smoother and more workable than what is used now. Modern-day fly ash is leftover waste from coal-fired power plants. More than 71 millions tons of fly ash were produced last year, and most of it ended up in the air or dumped into landfills, but it can replace more than half of the Portland cement in concrete mix.

Green $pecs

Overall Rating

Difficulty

Green Benefits

The technology used to make fly ash concrete dates back to the ancient Romans and was lost for centuries until it reappeared around 1914. It was first used in the U.S. in the construction of the Hoover Dam. It is poured and set just like traditional concrete mix.

What will this project do for your home?

The chemical reaction of fly ash with lime produces concrete with higher strength and durability than concrete made with Portland cement alone. The round spheres of fly ash also result in a smoother finish with fewer clumps and cracks. The higher density of fly–ash concrete requires less water when mixing and allows more time for setting. Fly–ash concrete also has fewer voids in the surface, making it more resistant to water damage and intrusion. Because of this benefit, many states now require high fly–ash content in concrete used for the construction of roads and bridges.

What will this project do for the Earth?

Portland cement production has an incredible impact on global energy use and carbon emissions. This industry alone has the ability to slow the effects of global warming and eliminate our dependency on foreign oil. Using fly ash in place of cement eliminates the negative impact on the planet, with every ton of fly ash used saving enough electricity to power a home for nearly a month.

Using fly ash in concrete also helps stop the ash from ending up in crowded landfills, where 65% of it now goes. Recycling this waste product into one home saves enough landfill space to contain the daily trash produced by 455 people.

Will you need a contractor?

Using fly ash in concrete for a new home or remodel requires one thing: communication. The contractor who mixes the concrete needs to know so he or she is prepared for slower drying time and increased workability. The engineer, who designs the structure of the concrete, needs to be prepared for the lighter weight and improved strength of the fly ash. Finally, the concrete supplier will have to be willing to add fly ash into the mix. The supplier could be resistant if he or she has never used it before, so look for someone with experience working with fly ash. I don't recommend mixing concrete yourself.

What are the best sources for materials?

Portland cement suppliers have been sneaking fly ash into their mixes for years. Up to 15% of cement is made with fly ash because it stretches the supply of cement and lowers production costs. Fly ash comes from the soot of coal-fired power plants, but not just any soot can be used. The ash at the bottom of the stack (called bottom ash) cannot be used in concrete. Avoid concrete made with fly ash from plants burning municipal solid waste, hazardous waste, medical waste, or tires. This leads to impurities in the fly ash that affect concrete performance. Look for concrete that uses at least 35% fly ash, but you could go as high as 50% with the approval of your structural engineer and contractor.

Fly ash is the leftover soot from coal-fired power plants that is usually dumped into the air or landfills. Under a microscope, fly ash is a series of round spheres that give concrete a smoother surface and stronger finish than if using Portland cement.

For any concrete that will be left exposed, ask the contractor to do a test piece to ensure color and finish are as desired. Many contractors using concrete with high levels of fly ash for the first time comment on how differently it responds to finishing. Any hidden concrete used in the foundation, slabs, or columns can be poured just like traditional concrete. The longer setting time may be an issue in instances where removal of concrete formwork is preventing other work from being done. Fly–ash concrete also requires less water than traditional concrete mix, so it's easier to pour in cold weather.

Fly ash contains trace amounts of heavy metals, including mercury and lead. There has been debate and confusion about the potential health effects of using fly ash, but recent studies from the Portland Cement Association have shown that concrete containing fly ash does not release any mercury or lead into the air and in fact absorbs the toxins.

Opponents of fly ash suggest its use promotes coal and coal-fired power plants. But since only 6% to 7% of the fly ash in the world is used, the demand is not really driving consumption. When the day comes that we start burning coal just to get fly ash, that's when we should reconsider using it.

How much maintenance will be required after installation?

Efflorescence is the white, powdery residue that appears on the surface of concrete after drying. It is evaporated salts leaching out of the surface and is harmless, but it can be removed with a light mix of muriatic acid and water.

RESOURCES

Fly ash suppliers:

American Coal Ash Association
www.acaa-usa.org

Boral Material Technologies Inc.
www.boralmti.com

Full Circle Solutions, Inc.
www.fcsi.biz

Headwaters Resources, Inc.
www.flyash.com

Lafarge North America
www.lafargenorthamerica.com

Nebraska Ash Co.
www.nebraskaash.com

The SEFA Group
www.sefagroup.com

Trans-Ash
www.transash.com

How long will the project take to accomplish?

Adding fly ash into concrete mixture does not add any additional time to a construction job. But you will notice a longer setting time for the concrete; plan ahead for this to avoid delays.

KEVIN SAYS Using fly ash in concrete is probably one of the easiest decisions to make. It does not affect the performance of concrete, helps the environment, and is financially neutral. There have been concerns that concrete mixed with fly ash may harbor heavy metal content left over from the coal-burning process, but most experts agree that almost all of those metals are left behind and that using fly ash in concrete is a safer way to dispose of it than dumping it in a landfill.

What is the capital cost?

After contacting several concrete suppliers, I learned that there is zero cost or savings associated with the use of fly ash. The concrete suppliers were supportive of its use and confirmed that it is abundant, beneficial, and has no impact on price.

What financial resources are available?

There are no tax credits or rebates associated with fly ash.

What is the monthly cost or savings?

There is no monthly cost or savings to be considered in using fly ash.

What is the long-term home value?

When it comes time to sell a home, it's likely no one will ask if fly ash is in the concrete, therefore there is zero impact on long-term value.

THE BOTTOM LINE is...

This is as easy as saying to your contractor or concrete provider, "I want 35% fly ash." Using fly ash has zero effect on you and plenty of positive impact on the planet. Go for it. ■

Incorporate advanced framing techniques

ERIC SAYS At the height of the building boom over 1.5 million homes were being built in the U.S. every year. The majority of them are constructed out of lengths of wood using a method called platform framing, which dates back 175 years and has changed very little over the decades. Back in the 1830s, platform framing was introduced as the Industrial Revolution brought about new methods of mass production. Standardized sizes of lumber and fasteners combined with platform framing methods allowed for a great expansion of housing across the U.S. It gained widespread acceptance in the 1940s, when the low skill and high speed of the system was ideal for the housing boom after World War II. Initially hailed as an efficient method of creating homes, wood framing now takes an incredible toll on the environment by rapidly accelerating deforestation.

Building a typical 2,000-sq.-ft. home consumes an acre of forest and more than 13,000 bd. ft. of framing lumber. End to end, the boards would extend nearly 2.5 miles. Adding up all the boards that go into homes built in just one year would extend to the moon and back almost 20 times. By using less wood to build our homes, we can preserve more of our forests, slow the effects of global warming, and protect the environment for future generations. But how can you build the same size house with less wood?

This is the basic principle behind advanced framing, an approach to platform framing that uses less wood, smarter joints, and fewer connections. Often referred to as optimum value engineering (OVE), advanced framing can cut the amount of wood you'd normally use to build a home by as much as half, just by framing in a more efficient way. And all of this can be done without changing a home's appearance or design.

Green $pecs

Overall Rating

Difficulty

Green Benefits

What will this project do for your home?

In total, advanced framing techniques can eliminate 30% to 50% of the amount of framing lumber that would normally go into a building. By using less wood, the structure is lighter and therefore stronger. Advanced framing reduces the strain on each structural member and connection by lining up the wall structure with the roof structure. Advanced framing also eliminates the uninsulated corners that are created with standard framing, and the thicker walls allow more room for insulation, which can help make a home more energy efficient.

What will this project do for the Earth?

Wood makes up nearly 40% of the construction waste generated from building a new home. Lumber harvesting and production is the second largest source of greenhouse gas emissions (after coal-fired energy production). Drastically cutting the amount of wood used to build a home would curb the 65 million tons of construction waste that gets sent directly to our crowded landfills each year. In the construction of just one house, advanced framing techniques prevent the destruction of 64 trees, or the equivalent of 12 tons of carbon emissions.

Will you need a contractor?

Only an experienced carpenter or contractor should build a home with wood. Building codes and structural requirements are not to be taken lightly, and the work should be left to the professionals. The techniques used in advanced framing methods require an understanding and knowledge of basic framing since advanced framing is really just an improvement on traditional wood-framing methods.

The use of advanced framing should be planned early in the design process to take advantage of potential savings. If building a new home, your architect and structural engineer are critical to maximizing the benefits of advanced framing. You may encounter resistance, since most architects and engineers reuse standard specifications for wood framing. Using advanced framing techniques may require them to revise standard drawings. Seek out a building team familiar with OVE. Many building inspectors unfamiliar with OVE will raise a flag of caution, but OVE is accepted by all three national building codes. Once you have an approved permit to use OVE, no building inspector can reject the use of it.

What are the best sources for materials?

In standard wood framing, wood studs are spaced 16 in. apart and nailed into place to create a wall. The reason for this spacing is because building materials are based on increments of 8 in. Bricks are 8 in. long; concrete blocks are 16 in. long; plywood is 48 in. long. The original intention with framing was to design buildings using multiples of 8 in. and, as a result, reduce the amount of cuts made, produce less waste, and speed up the entire construction process. But many homes are made using arbitrary dimensions (39 ft.; 2¹/₂ in.). Taking advantage of the 8-in. module is the central idea of advanced framing.

Advanced framing techniques use the same materials as traditional platform framing techniques. Framing wood, connectors, and fasteners can be purchased at most any lumberyard or home improvement store. The greenest way to build with wood is to purchase wood certified by the Forest Stewardship Council (FSC) (see p. 235) or use engineered wood rather than solid pieces of wood for framing. Made up of small scraps of wood held together with an industrial adhesive, engineered wood is stronger and straighter than natural wood and allows smaller dimensions to be used (see p. 240).

There's a misconception that using less wood used in advanced framing makes a home weaker. On the contrary, a lighter structure

(continued on p. 224)

The construction of a typical home consumes an acre of forest. The 175-year-old practice of wood framing changed very little over the years—causing extensive deforestation—until advanced framing was developed, which uses less wood to accomplish the same task.

The Eight Concepts of Advanced Framing

Advanced framing techniques are founded
on eight ideas. Added together, these
approaches can result in a 30% to
50% reduction in the amount of wood
used in a new home.

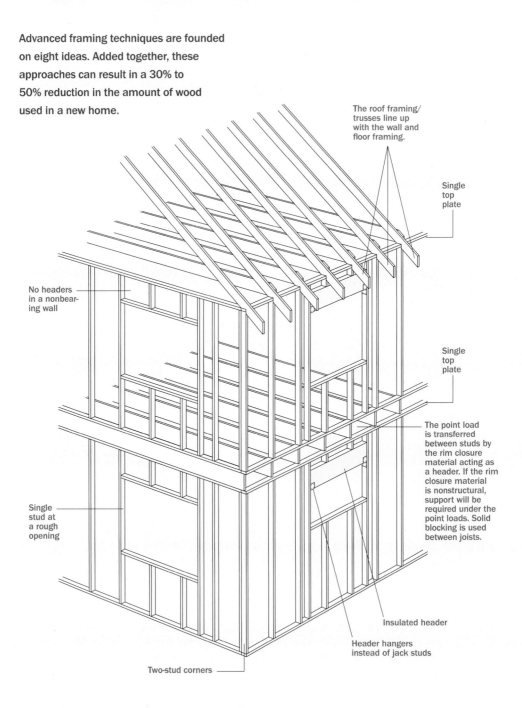

The roof framing/
trusses line up
with the wall and
floor framing.

Single
top
plate

No headers
in a nonbear-
ing wall

Single
top
plate

The point load
is transferred
between studs by
the rim closure
material acting as
a header. If the rim
closure material
is nonstructural,
support will be
required under the
point loads. Solid
blocking is used
between joists.

Single
stud at
a rough
opening

Insulated header

Header hangers
instead of jack studs

Two-stud corners

1 **Framing spacing of 24 in.** If 16 in. works with an 8-in. module, so does 24 in. Advanced framing spaces wall, roof, and floor framing at 24 in., which reduces the amount of wood (and labor) that goes into a wall by one-third. It also speeds up construction time and ensures stronger structural connections.

2 **Thicker walls.** If a standard wall utilizes 2-in. × 4-in. wood studs, advanced framing uses 2-in. × 6-in. studs. The extra 2 in. of depth not only add strength but also provide additional space for insulation.

3 **Single top plate.** Since roof and floor members line up to wall members, only one plate is needed at the top of each wall. Called the top plate, it distributes the weight of the roof into the wall. It is held together with simple metal connector plates, reducing weight and speeding up construction.

4 **Uninterrupted spacing.** Windows and doors should be designed to fit within the 24-in. module. Otherwise, they slow down the construction process and increase waste. By paying attention to the location of every wood stud, one can be better informed in designing the layout of the windows and doors.

5 **Bridge at intersecting walls.** When an interior wall intersects an exterior wall, a small piece of wall (called blocking) is inserted between the studs to create support. Metal clips could also be used at the top and bottom of the interior wall to hold it in place.

6 **Modular layout.** A house should be designed to take advantage of 24-in. modular spacing. This eliminates the need to cut up scraps of wood, plywood, insulation, and siding—all of which come in standard dimensions of 24 in.

7 **Right-sized headers.** A header is the beam that supports a wall over a window or door. Most of the time, headers are all made the same size, but have your architect or structural engineer size each header appropriately. This reduces the size of most of the headers to save wood.

8 **Open corners.** Rather than inserting an extra stud into a corner to hold two walls together, metal connectors and clips should be used so that the corners remain clear to receive insulation, making the building more energy efficient.

(continued from p. 221)

with fewer stressful connections makes a home stronger. There are no state-specific bans on using OVE. In spite of this, some code officials unfamiliar with OVE may be reluctant to approve your permit. You may need to educate them with helpful articles from Building Science Corporation (www.buildingscienceconsulting.com) or Toolbase (www.toolbase.org).

How much maintenance will be required after installation?

Since fewer but wider studs are used to create an OVE wall there's room for more insulation, making it more energy efficient. Once complete, advanced framing requires no more or less maintenance than traditional wall framing.

How long will the project take to accomplish?

By design, OVE requires fewer boards and studs to install. Experienced contractors will be able to frame using OVE techniques in one-third less time than that required for traditional framing. For contractors unfamiliar with OVE techniques, using advanced framing may add time to the initial staging and planning of a project.

KEVIN SAYS The good news about advanced framing techniques is that, done right, it not only helps the planet but also saves money immediately. The challenge is the limitations it imposes on design as well as geographic restrictions for safety. Several engineers who specialize in green building tout OVE as an excellent approach for simple production homebuilding. However, more than one engineer interviewed pointed out that it often doesn't apply to custom homes because certain features and designs may require more extensive framing not achievable through OVE. Then there is the challenge of finding labor that can execute the techniques. Many contractors and framers are uncomfortable attempting and warranting unfamiliar techniques, which may limit your choices in picking a building team.

What is the capital cost? $$

With the reduction in wood and framing labor, estimates from the Department of Energy and the National Association of Home Builders (NAHB) put savings at roughly $1 per sq. ft. of building. Green associations such as the National Resources Defense Council (NRDC) estimate even higher savings. Assuming a custom home design allowed for OVE, a 2,000-sq.-ft. house would save $2,000 on the build. The savings requires the use of framers who are familiar with OVE techniques since the labor required may not be less if they have to learn a new process.

What financial resources are available?

For consumers building a custom home or remodeling an existing house, we couldn't find any tax breaks or rebates for the use of advanced framing techniques.

What is the monthly cost or savings? $$$$

The Department of Energy and NAHB have both done recent studies showing that the reduced lumber used in OVE allows for more insulated area, resulting in more efficient heating and cooling. For a home with energy bills of $100 per month, you can expect a monthly savings of roughly $20 to $30.

What is the long-term home value?

The major resale factor related to OVE will be the energy savings. While there is no specific data to support a higher sales price for OVE, the energy savings may be attractive to a buyer. Since energy bills will be established at the completion of the house, savings will be difficult to demonstrate since there is no direct comparison. In earthquake- and hurricane-prone states, the technique may raise unsubstantiated safety concerns, making the home more difficult to sell.

THE BOTTOM LINE IS...

If OVE methods are appropriate for your location and will work for the design of your home, then this approach is a hands-down winner with instant and long-term savings. ■

RESOURCES

Building Science Corporation
www.buildingscience
consulting.com

Energy & Environmental Building Alliance (EEBA)
www.eeba.org

Toolbase
www.toolbase.org

Build with reclaimed materials

Green $pecs

Overall Rating

Difficulty

Green Benefits

ERIC SAYS Valuable resources such as wood, metal, and stone are thrown away every day in the construction industry despite being in good condition and reusable. Meanwhile, most natural resources used to make these materials are being exhausted and destroyed. Reclaiming old materials for new use is a simple, effective way to furbish a home that has little to no negative environmental impact and provides an opportunity to capture the quality and charm of old building products.

Although many manufacturers offer products with recycled content, using recycled materials is not always the greenest option. It requires a lot of energy to recycle materials not originally designed to be recycled, like steel and aluminum. Although recycling unwanted or unused items is better than producing virgin (new) items, reclaiming materials is ideal because it avoids the energy-intensive processing of recycling. Instead, they are salvaged, cleaned up, and used over again.

What will this project do for your home?

Reclaimed materials offer practical benefits over new products. You can often find better-quality materials than are currently available, as in the case of stone. Reclaimed wood has often dried out and adjusted to its environment, making it less likely to warp or twist. Reclaimed materials also offer homeowners the opportunity to insert a piece of history and one-of-a-kind character into the home, as in the case of an old claw-foot tub or antique fireplace mantel.

What will this project do for the Earth?

By reusing materials intended to be thrown away, you save them from taking up space in a landfill. And if you're reusing a material that

LEFT Recycled wood can be milled and used to furbish homes with quality old-growth lumber for timbers, flooring, and finish materials.

RIGHT Wood can be reclaimed from a variety of sources. This old barn could easily be deconstructed to salvage the wood that can be made into flooring.

already exists, energy that normally would have gone into making a new product is also saved.

Will you need a contractor?

The use of reclaimed materials takes a little research but is nothing you can't do yourself. Ask your architect or contractor about finding architectural salvage yards. Locating a wood floor made from reclaimed wood may be easy, but finding a source for reclaimed brick might take more effort. If buying reclaimed cabinets (or anything that needs to fit in a specific location), take the time to measure before buying.

What are the best sources for materials?

Start by locating architectural salvage, building material salvage, and material exchange companies in your area. If there are no local resources, contact a local demolition contractor for possible leads on reclaimed and salvaged materials.

Other local sources depend on the material. Metal yards sort and store scrap pieces that can be had for little or no cost. Granite and marble countertop showrooms often have boneyards where leftover slabs of stone are kept. Tile stores can't sell any box of tiles that has been dropped. You might be able to pick up the discards for free.

A list of salvage yards can be found at the Architectural Salvage Exchange (http://salvageweb.com). The Reuse People (www.thereuse people.org) and Habitat for Humanity (www.habitat.org) may be able to provide salvaged and reclaimed materials as well, and the Building Materials Reuse Association (www.bmra.org) may offer a listing of materials in your area.

How much maintenance will be required after installation?

The amount of maintenance varies based on the material and condition of the reclaimed item. In general, a reclaimed material would not need any more maintenance than a new product.

How long will the project take to accomplish?

Sourcing reclaimed materials may add time to the initial design process, but installation will likely be the same as when putting in a new material.

KEVIN SAYS Some of the most beautiful and unique materials I have seen in homes have been reclaimed. Although it would seem like these materials would be less expensive, there is additional cost in preparing materials for reuse including treatment for infestation and weatherizing. But scouring salvage yards can be fun and may yield some truly interesting finds at good prices.

Reclaiming Materials

Like any antique hunt, you may have to visit a salvage yard several times before finding what you want. It helps if you visit with a shopping list so you know ahead of time what to look for and the sizes you need. Bring along a tape measure and camera, too.

Wood: Reclaimed wood sources vary from old barns to lost wood that has been submerged underwater for decades. It can be made into flooring, siding, trim, and cabinetry. Larger pieces can be fashioned into decorative beams and trusses for architectural features.

Brick: Old discarded bricks can be reused as brick pavers and flooring. Reclaimed brick has a distinctive, weathered look and irregular color.

Stone: Any and every type of stone is available from reclaimed sources. Sometimes it is left over from partially used full slabs. Other times the stone has simply been discarded. More than 526 million sq. ft. of granite countertops were sold last year, replacing older countertops that can be reclaimed. Slate and clay roofing tiles can also be readily found.

Fixtures: Cabinets, doors, and light fixtures are easy to salvage and reuse, but confirm the size you need before buying. Old windows can also be reclaimed, but it's not recommended. They may not be as energy efficient as new windows and end up wasting more energy in the long run.

Old plumbing fixtures such as tubs, sinks, and even toilets can be reclaimed, and enamel fixtures can be refinished to look like new. If buying a salvaged toilet, make sure it is a low-flow model (see p. 13).

Test any painted items for lead paint. Test kits are available at hardware and home-supply stores. If old wood contains lead paint, it must be stripped and sealed before it can be used.

What is the capital cost? $$

The cost of reclaimed materials is generally a bit higher than for new materials. For example, new hardwood flooring runs anywhere from $3 to $12 per sq. ft. For a 2,000-sq.-ft. home, flooring materials would cost $6,000 to $24,000 (and up to $40,000 for exotic hardwoods). Reclaimed options stretch from $8 to $20 per sq. ft., costing $16,000 to $40,000. Labor and installation for new and reclaimed items are generally the same.

What financial resources are available? $

Tax credits are few and far between. There are some small tax credits in states such as Maine and Montana for using reclaimed materials, and you can save 5% to 10% at salvage yards by offering cash deals and bartering, but that's about it.

What is the monthly cost or savings?

There is no direct monthly cost or savings impact from using reclaimed materials.

What is the long-term home value? $$

The use of quality finish materials always has a positive impact on home resale value when tastefully done. The resale value does not, however, correspond dollar-for-dollar. Appraisers often give about 70% of cost as a credit for quality improvements, but this percentage diminishes as cost gets higher. There is no data to show reclaimed materials sell a home for more than one furbished with new materials.

THE BOTTOM LINE IS...

If you can afford the extra cost, reclaimed materials can provide unique beauty to a home and help the planet as well. Finances won't be the primary motivator here, but you'll feel better about your green contribution. ▮

RESOURCES

Aged Woods®
www.agedwoods.com

AquaTimber™
www.aquatimber.com

Architectural Salvage Exchange
http://salvageweb.com

Architectural Timber and Millwork
www.atimber.com

Barnstormers
www.barnstormersinc.com

Biotimber™
www.biotimber.com

Black's Farmwood
www.blacksfarmwood.com

The Building Materials Reuse Association
www.bmra.org

Carlisle Restoration Lumber
www.wideplankflooring.com

Crossroads Recycled Lumber
www.crossroadslumber.com

EcoTimber
www.ecotimber.com

Mountain Lumber Co.
www.mountainlumber.com

Pioneer Millworks
www.pioneermillworks.com

Restoration Timber
www.restorationtimber.com

Terra Mai
www.terramai.com

Timeless Timber
www.timelesstimber.com

Trestlewood®
www.trestlewood.com

Build with recycled drywall

ERIC SAYS Drywall is the standard interior finish for homes around the world. The average new home requires more than 8 tons of it, but about 12% of that is thrown away, with cut pieces, scraps, and leftover boards contributing to a considerable waste problem. Approximately 15 million tons of drywall are produced annually in the U.S. to keep up with the demand. It goes by many names—Sheetrock®, gypsum wallboard, or plaster-board. By any name, however, drywall takes a terrible toll on our environment, from production through disposal.

Invented in 1916 as a cheap and less labor-intensive replacement for plaster, drywall is made with raw gypsum (calcium sulfate) mined from the ground and heated to treat and stabilize it. Then the gypsum is ground up and mixed with several additives to provide fire, water, and mold resistance. The wet gypsum mix is formed into a panel and sandwiched between two layers of heavy, recycled paper. The panel is then heated in a large kiln, where the sandwich hardens and becomes strong enough to be used as a wall panel.

All of this processing totals up to a large bill for the earth. Mining raw gypsum destroys the surrounding environment. Heating the raw mix and firing the panels consumes almost 1% of all U.S. energy, and manufacturing drywall produces 51 million tons of greenhouse gases every year.

The disposal of drywall is another problem. Although drywall is a standard part of any building, it is also one of the materials least likely to get recycled. Of the 164 million tons of construction waste produced every year, a quarter of that is drywall. Nearly 10% of the contents of every landfill is scrap drywall and more than 41 million tons get tossed into crowded landfills every year. While in the landfill, the gypsum in drywall scraps breaks down and releases hydrogen sulfide gas. As a result, some landfills refuse to take drywall. Another disposal option,

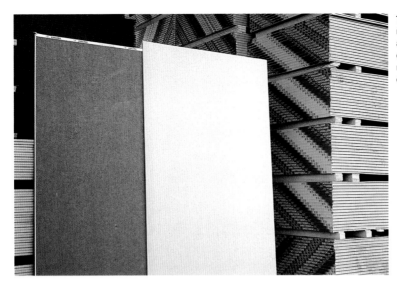

The production of drywall requires mining, heating, mixing, and firing. All of this embodied energy takes a toll on the environment, but that can be avoided by using recycled drywall.

incineration, is also problematic because drywall scraps produce sulfur dioxide gas when burned. (Both gases are toxic to humans.) By using recycled or synthetic drywall, you can help eliminate the intense energy use and disposal issues associated with drywall.

What will this project do for your home?

By switching to recycled or synthetic gypsum, you will not notice any difference in the installation, finishing, or maintenance of the wall. The only difference is that you'll have made a smarter, more environmentally responsible choice.

What will this project do for the Earth?

Producing 1,000 sq. ft. of drywall requires around 2 million Btu of natural gas and generates 234 lb. of carbon dioxide. Eliminating the greenhouse effects of drywall production alone could cut carbon emissions to the level scientists say we need to reach by the year 2020. Just this one industry alone could solve the global-warming crisis.

Will you need a contractor to complete this project?

Drywall can be installed by almost anyone. To ensure you are getting recycled drywall, it must be specified by the architect and purchased by the contractor. Although the paper facing on drywall is almost exclusively made from recycled content, you cannot assume the cores of the boards are also recycled. Specifically request and order recycled-content boards just to be sure.

The percentage of recycled content varies from manufacturer to manufacturer and batch to batch. Be sure to ask for the exact amount of recycled content in your boards. Boards made of 100% recycled content are available.

What are the best sources for materials?

Drywall is sold at home improvement stores and lumberyards around the country. A large retailer should carry recycled and synthetic drywall. If not, it can be ordered and delivered.

With careful planning, the amount of wasted drywall in any project can be reduced. Most drywall comes in a standard width of 48 in. and standard lengths of 8 ft. and 9 ft. Designing a ceiling height of exactly 8 ft. or 9 ft. eliminates the need to cut strips of drywall to fill in the last bit on an 8-ft. 2-in. wall, for example.

Again, nearly every drywall manufacturer uses recycled paper facing on its products. This does not make it a green product despite any claims a manufacturer may make. Look for recycled or synthetic gyp-

The demolition of a typical home produces one ton of drywall waste. In total, drywall makes up one-quarter of all construction waste.

sum cores, and instead of paper facing, which can become a medium for mold growth, look for paperless boards (reinforced with fiberglass or cellulose) for damp or wet areas such as basements and bathrooms.

Synthetic gypsum In an effort to cut sulfur dioxide emissions from coal-fired power plants, devices called scrubbers are installed to remove the pollution and prevent it from entering the air. Scrubbers use a method called flue gas desulfurization (FGD) to scrub the emissions. FGD produces synthetic gypsum as a by-product that can be used to replace gypsum and create 100% synthetic drywall panels.

FGD panels are available only in certain areas. If you have to ship it from long distances, it will ultimately produce more pollution than it prevents. Only use it if you can find it within 500 miles of your home.

Recycled-content gypsum Given the large amount of drywall waste produced each year, recycled gypsum is easy to source and use. The amount of recycled content varies by product and could be anywhere from 5% to 100%. Simply seeing the word recycled is not enough; ask how high that percentage really is.

How much maintenance will be required after installation?

Recycled or synthetic drywall requires no additional care or maintenance. It can be finished and painted just like standard drywall.

How long will the project take to accomplish?

You may have to special-order recycled drywall if your local home improvement store doesn't keep it in stock. Call at least eight weeks before you need the drywall for your project.

Unlike the lath and plaster techniques used in homes before World War II, drywall can be installed in just a couple of days. Switching over to recycled-content drywall will not add any additional installation time.

KEVIN SAYS Drywall is one of those items that we have around us all day but don't even think about. We stare at our walls without much thought for what goes into them. The idea of recycled gypsum in drywall is so convenient and beneficial that it makes you wonder why new is even an option. Finding recycled-content drywall at the local big-box stores requires a bit of looking, but luckily there are plenty of online suppliers. And if demand for recycled-content drywall picks up, it is likely that big-box stores will start pushing planet-helping options more aggressively.

What is the capital cost?

The good news is that drywall with 100% recycled gypsum content—while a little tough to find—is about the same cost as new drywall. Individual sheets of 4-ft. × 8-ft. drywall typically cost about $10 each. Installation cost would be the same for both new and recycled material. This means a room that is 16 ft. × 16 ft. with 12-ft. ceilings will need about $200 to $240 in drywall material. A 3,000-sq.-ft. house might run as much as $3,000 to $5,000 in total drywall material.

What financial resources are available?

There do not appear to be any specific tax credits or rebates for the use of recycled drywall material.

What is the monthly cost or savings?

There is zero monthly cost or savings impact associated with using recycled drywall material.

What is the long-term home value?

There is no long-term value impact from using recycled or synthetic drywall material.

THE BOTTOM LINE is...

It's tough to find an argument against using drywall with recycled content. It is cost neutral, helps the planet, and has no downside in terms of the value of your property. Insist that your contractor makes this happen—preferably using 100% recycled drywall. Skip the synthetics until they become price competitive. ▓

Build with certified wood

ERIC SAYS An area of forest the size of a football field is cut down every two seconds. In the span of an entire year, 32 million acres of forest are lost forever. That's a region larger than the entire state of North Carolina destroyed every year. In the U.S. alone, 96% of old-growth forests have been destroyed. Worldwide, 80% of ancient forests have been removed. The small amount remaining is under direct threat from logging and development, much of it under illegal practices.

Forests provide many important and valuable services for free. They stabilize soil, regulate climate, and pump humidity into the air. Trees filter and absorb carbon (in the form of carbon dioxide), helping to slow the threat of global warming. And let's not forget that trees produce oxygen for us to breathe.

The economic value of a forest doesn't come from pulling wood out of the forest, but keeping wood *in* the forest. Scientists estimate the service values that global forests provide (air, soil, carbon filtering) to be around $4.7 trillion a year. In the near future, the carbon savings from maintaining forests will be more valuable than the wood harvested by cutting it down.

In fact, destroying forests costs the global economy even more. Nearly 25% of all carbon emissions are a by-product of deforestation. The boreal, tropical, and temperate forests of the world store 1,200 gigatons (that's billions of tons) of carbon in their roots and surrounding soils. The 51 million urban trees within the forests of Chicago alone store approximately 155,000 tons of carbon. When cut down, dead trees release more stored carbon than the annual emissions from all cars and trucks in the U.S.

A large percentage of logged wood goes into construction (lumber, paneling, cabinets, and furniture), and the $270 billion-a-year lumber industry drives consumption up every year. At this rate, the world's

Green $pecs

Overall Rating

Difficulty

Green Benefits

Founded in 1993, the FSC has emerged as the trusted leader in certified wood. It sets standards for what constitutes responsible forest management and certifies organizations that follow those standards.

rainforests will vanish by the end of this century. The obvious cure for deforestation is to stop cutting down trees, or at the very least, manage forests so they last. That is the idea behind the Forest Stewardship Council (FSC). The FSC is an international nonprofit organization that certifies wood has been sustainably harvested from well-managed forests. The FSC stamp is a guarantee that the wood you use was not clear-cut (leaving bare land) and its harvesting did not contribute to deforestation. Any product made from wood can now be created from an FSC-certified source.

What will this project do for your home?

Building a home requires approximately an acre of wood, for framing, floors, siding, cabinets, doors, trim, and paneling. Other than how the wood was sourced, there is no physical difference between FSC-certified studs and those made from a clear-cut forest. That's why convincing ourselves of the need to use only FSC-certified wood can be difficult. The results of the extra money we've spent for certified wood look the same and function identically to any other wood. We can rest easier knowing that the acre of lumber for each of our homes did not contribute to deforestation or global warming, and maybe tell guests about a commitment to sustainability. Otherwise, I agree—it's a hard pill to swallow.

But as more people choose certified wood and more manufacturers use it exclusively, the market will shift. In the future—whether by choice or regulation—certified wood will be the only wood available. In the meantime, selecting FSC-certified wood is one of the most impactful ways to green a new home.

What will this project do for the Earth?

Choosing FSC-certified wood helps reduce illegal logging practices and shifts the lumber industry toward more sustainable practices. Certified wood use ensures there will be trees to absorb carbon, filter the air, and stabilize the climate. Using certified wood prevents further deforestation and averts the release of stored carbon into the atmosphere.

Will you need a contractor?

If you want FSC-certified wood in your new home project, you'll need to communicate that to your contractor. Since the contractor does most of the purchasing, he or she needs to understand how important this choice is to you. Some certified-wood products may have to be specially ordered or requested, so research and plan ahead to avoid building delays.

What are the best sources for materials?

Certifying a product is intended as a type of seal of approval. The FSC stamp alerts consumers that a wood product comes from a forest managed under strict environmental guidelines.

National retailers, such as The Home Depot (www.homedepot.com) and Lowe's (www.lowes.com), offer limited supplies of FSC-certified wood products. The FSC has an online product directory (www.findfsc.org) and will help locate products if you send product requests to info@fscus.org. Be detailed in your e-mail request.

To verify the claims of a manufacturer regarding certified wood, ask for a copy of the chain of custody, a document that tracks wood from initial logging through manufacturing and distribution. The tracking information can then be verified on the FSC website (www.fscus.org). The chain of custody is the only proof you have that FSC standards were followed.

The success of the FSC has created demand for other certification systems around the world. Unfortunately, most have been created by the lumber industry itself to provide a lower standard more favorable to existing practices. The Sustainable Forestry Initiative (SFI) is a competing set of guidelines developed by the American Forest and Paper

Construction of a typical home requires 13,000 bd. ft. of lumber. Siding, flooring, and trim add an additional 8,000 sq. ft. Making the decision to use FSC-certified products has a positive effect on the environment, with absolutely no difference in the quality and structural soundness of a home.

Wood Wise

- More than 13,000 bd. ft. of lumber go into constructing a new home. Switching to FSC-certified wood prevents the clear-cutting of an entire acre of forest.

- More than 6,200 sq. ft. of plywood goes into finishing the walls of a home. Switching all of the plywood, strand board, and other wood panels to certified wood protects our forests and the strength of the home is not diminished at all.

- Trim is used throughout a home around windows, doors, and along the wall base to highlight and accentuate edges. Select finger-jointed trim, made

up of small pieces of wood stitched together into one long piece. This saves large wood from being used and takes advantage of discarded scrap material.

- Technically, any wood door manufacturer should be able to substitute certified wood into its products. An average home has around 20 doors, and most of these can be made from certified wood.

- Any custom cabinetmaker could switch to certified wood upon request. Certified wood is not limited in wood color or species available for your cabinets.

Association, an industry trade group. SFI is voluntary and self-enforced (with no oversight). Arguably, it was created by the lumber industry to greenwash itself and deflect environmental criticism. Despite shortcomings, many lumber manufacturers push the SFI standards, but SFI is not an equivalent substitute for FSC standards. An informational website has been created by a group of environmental agencies to educate consumers about the inadequacies of SFI certification (www.dontbuysfi.com). To date, the FSC certification remains the only credible and trusted seal for sustainably harvested wood.

How much maintenance will be required after installation?

Wood certified by the FSC requires no more and no less maintenance than any clear-cut wood product.

How long will the project take to accomplish?

If research and ordering FSC-certified wood is done in the early planning and design phases, there should be no additional time added to the overall length of a project.

KEVIN SAYS Using wood certified as ecologically friendly is a worthwhile endeavor. Execution of the concept, however, can be a bit challenging. Since FSC certification focuses on managing the entire supply chain, you need to find a retailer that has access to an FSC wholesaler. The FSC only lists about 50 retailers nationwide, with the largest concentration of them in northern California. The rest are sparsely spread out across the country, and there's not one in every state. Big-box stores are set up to sell FSC products, but they are not identified online, and in my store visits, employees were not able to identify products that were FSC certified.

After calling several contractors, lumber suppliers, and distributors, I learned that FSC-certified wood currently accounts for less than 20% of sales at FSC-certified retailers. Often consumers will inquire but balk at the premium in price. Many retailers don't want to pay the premium to the FSC for a low-demand product, which makes it hard to acquire.

Contractors are unlikely to request it since it makes bids higher, so using FSC-certified products demands extra fortitude on the part of the consumer.

What is the capital cost? $$$

Costs of lumber and trim vary greatly from home to home, but the difference in cost for comparable FSC products is about 20% higher across the board. Some of the cost is associated with the actual management of the forests, but much of it comes from the additional monitoring and recordkeeping by the manufacturers and distributors as well as fees paid to the FSC. Lumber suppliers and retailers claim to carry the same markup on FSC products as they do non-FSC lumber. So if a 2,000-sq.-ft. house requires $30,000 in lumber, going with an FSC product will cost you an additional $6,000 or so.

What financial resources are available?

Although commercial builders can benefit from using LEED techniques, there do not appear to be any specific tax credits or rebates for the use of FSC wood by consumers.

What is the monthly cost or savings?

There is zero monthly cost or savings impact associated with using FSC wood.

What is the long-term home value?

There is no specific long-term value impact for using FSC wood, but an ecologically minded buyer might choose your home over an alternative built without FSC-certified wood.

THE BOTTOM LINE IS...

Using FSC-certified wood is an honorable idea worth considering, provided you can afford the premium and live near a supplier. However, the price and hassle of finding materials could be a major deterrent to both you and your contractor. ■

RESOURCES

Forest Stewardship Council (FSC)
www.fscus.org

FSC Product Directory
www.findfsc.org

SFI compared to FSC Standards
www.dontbuysfi.com

Doors and windows:

Algoma Hardwoods
www.algomahardwoods.com

Eggers Industries
www.eggersindustries.com

VT Industries
www.vtindustries.com

Furniture:

Berkeley Mills
www.berkeleymills.com

Environmental Language
www.el-furniture.com

Trim:

Architectural Millwork Mfg. Co.
www.archmillwork.com

The Collins Companies
www.collinswood.com

Wall covering and paneling:

CraftWood®
www.srwoodinc.com

EarthSource Forest Products®
www.earthsourcewood.com

Build with engineered lumber

Green $pecs

Overall Rating

Difficulty

Green Benefits

ERIC SAYS Wood is relatively strong, durable—yet easily cut and drilled—and a highly adaptable construction material. When homebuilding was at its peak, more than 1.5 million homes were built out of wood in the U.S. each year, and much of the lumber had to be large, solid pieces.

Large pieces of wood must be harvested from large trees. Unfortunately, global deforestation has reduced the supply of large-diameter and old-growth trees. But engineered wood products offer a solution. Made from small trees and scrap bits of wood, engineered lumber is produced by compressing small pieces of wood together with an adhesive to create lumber that uses half the wood but has twice the strength of solid-wood materials.

Engineered lumber has grown in popularity due to technological advancements and the high price of solid wood and steel, and now floor joists, I-joists, posts, columns, beams, and sheet products can all be engineered from scraps compressed with adhesive. Switching all structural framing in a building project from solid wood to engineered lumber slows the rate of deforestation and builds a stronger home at the same time.

What will this project do for your home?

Engineered lumber offers numerous benefits over solid wood. The compressed members are stronger, more durable, and more termite resistant. Unlike regular wood, engineered wood won't shrink, warp, or twist due to humidity or natural knots in the grain. Once installed, floors built from engineered wood tend to squeak less than those of solid wood. The predictability of the size and quality of engineered wood often makes it the preferred choice for builders.

Consistent size and quality make engineered wood a top choice for most builders. As a general rule, anytime a piece of wood larger than a 2×8 is needed, substitute an engineered product instead.

What will this project do for the Earth?

Engineered wood products save the forests of the Earth. Not only are they produced from smaller-diameter trees and scraps of wood, but the added strength of engineered lumber allows them to support the same weight with a smaller beam.

Switching the lumber in a new house to engineered wood could easily preserve 15 to 20 trees. Half-a-million acres of forest could be saved each year if every home in the U.S. used stronger and more efficient engineered lumber.

Will you need a contractor?

For an experienced carpenter, changing from solid lumber to engineered lumber should be simple. Most contractors are familiar with the use and installation of engineered products. Instead of solid-wood studs, engineered studs are used in their place, and some connectors will change. But like any wood product, the same nails and screws can be used, and working with engineered wood means faster installation.

The architect or structural engineer will need to know early in the design process about using engineered lumber extensively. They will have to adjust construction drawings and permit documents to reflect the change.

What are the best sources for materials?

The rise in popularity makes finding engineered products easy. Any lumberyard, home improvement store, or hardware store will stock a variety of engineered wood.

Engineered lumber falls into two main types: veneers and products. Veneers, such as plywood, are shaved off a whole log placed onto a roller (imagine the tree being peeled like an apple). The large sheets are cut to size and glued together to form a rigid panel. Engineered products typically piece together wood waste, scraps, and smaller-diameter trees to make a number of materials. The products typically have a scrappy appearance, and individual pieces of wood are often still visible.

Engineered veneer and boards

- To make plywood, thin sheets of solid wood are bonded together in alternating directions. Typically, an odd number of sheets (usually three or five) are used. This unique arrangement makes plywood incredibly strong, while the alternating direction of the wood grain keeps the panels from expanding and shrinking due to humidity. Plywood is often used to add earthquake and hurricane resistance to wood buildings. When buying plywood, avoid luan-based products. Luan (also referred to as lauan) is wood made from tropical

A glue laminated beam can support twice the weight in half the size of a solid-wood beam.

rainforest trees. Chosen for its attractive finish, luan is nearing extinction and its harvesting contributes to the devastation of global rainforests.

- Similar to plywood, oriented strand board (OSB) is made up of small pieces of various types of wood. OSB is nearly as strong as plywood, but its appearance is much different.
- Particleboard consists of compressed sawdust and adhesive. Although not as strong as plywood, it is suitable for nonstructural applications and has a sandy appearance.
- Strawboard uses compressed wheat or straw. The heat generated during manufacturing activates the natural starch in the straw to act as an adhesive. Strawboard is ideal for homeowners sensitive to the adhesives used in engineered products.

Engineered products

- An I-joist (often referred to as a truss joist) consists of two thick, solid lengths of wood along the top and bottom with a thin web of OSB between them. The lightweight joists are stronger than solid wood and can be spaced farther apart or allow for a narrower joist to be used. For example, instead of 12-in.-deep solid-wood joists used to build a floor, an 8-in.-deep I-joist can be used instead.
- Glue laminated beams (glulams) are long strands of wood bonded together in alternating directions to create a thick, solid beam. Similar to plywood, glulams gain their incredible strength from the odd number of alternating layers used. Glulams are ideal in locations where a solid-wood beam is not strong enough and a steel beam is too costly or complicated.
- Similar to glulams, laminated veneer lumber (LVL) is composed of thin strips to form a wood stud. Ideally, LVLs would be used to replace all of the studs in a house. Their manufactured size is consistently straight and exact.
- Oriented strand lumber (OSL) is similar to LVL, except it uses strands of wood instead of thin layers. OSL is similar to OSB, except OSL is made into studs and beams.
- Parallel strand lumber (PSL) is similar to OSL and the two are often used interchangeably. Its appearance is usually knotted.
- Finger-jointed studs are small pieces of solid wood that have been stitched together to form a continuous length of wood. They are an excellent use of scrap wood but are not accepted by every building official for structural uses. Check with your local building department or use finger-jointed studs only in non-load-bearing walls.

When selecting engineered lumber, look for lumber with no added formaldehyde. Formaldehyde is primarily used as the adhesive in engineered wood but is a known carcinogen. All wood naturally produces formaldehyde, but some manufacturers will inaccurately refer to their products as formaldehyde-free. If you must use a product using formaldehyde binders, choose phenol-formaldehyde (PF) over urea-formaldehyde (UF), as it is slightly less toxic.

Solid wood is not as dimensionally stable as engineered wood. If mixing solid and engineered wood together, make sure the contractor allows the solid wood to dry out for two weeks before installation. Otherwise, it may shrink and cause construction defects where solid meets engineered wood. Save scraps of leftover engineered wood because they can be easily reused for blocking and bracing.

How much maintenance will be required after installation?

Because of its resistance to humidity and consistency in size, engineered wood reduces the number of construction defects and callbacks. When installed correctly, the use of engineered wood should reduce the amount of routine maintenance required for wood buildings.

How long will the project take to accomplish?

The use of engineered lumber speeds up construction, although time savings will vary based on the project. Coordinating the use of engineered wood should be done in the early planning phases, although some contractors can switch to engineered products during construction and suffer no delays from building inspectors.

KEVIN SAYS Engineered lumber has been around for quite a while and has become the norm in both production and custom homebuilding. Even 15 years ago, I remember asking contractors about OSB. They told me they use it to prevent floors from squeaking. (I mentioned that floor squeaks were how I knew where everyone was in my 80-year-old house.) As engineered lumber has become popular for its practical uses, the costs have become competitive in the overall building budget.

What is the capital cost?

Using engineered lumber is cost neutral for the most part. Although OSB may be nominally more expensive than using dimensional wood planking for a subfloor, the practice has become standard for most builders and brings with it structural benefits that go beyond item-for-item cost. The use of glulams and engineered I-beams allows for less structural wood overall even though the individual beams may cost 15% or so more than a comparable-size piece of dimensional lumber.

What financial resources are available?

There do not appear to be any specific tax credits or rebates for the use of engineered lumber.

What is the monthly cost or savings?

There is zero monthly cost or savings impact associated with engineered lumber.

What is the long-term home value? $

Newer homes using OSB usually have floors that don't squeak. This may help retain a bit of value over a creaky house, although there is no specific data to support the theory. Other than that, there is no long-term value impact from using engineered lumber.

THE BOTTOM LINE IS...

Most likely a contractor will want to use engineered lumber in a home and it's likely an architect will design with it in mind. But it doesn't hurt to ask just to make sure—especially since it won't cost any more money and will benefit the Earth. ■

RESOURCES

Standard Structures
www.standardstructures.com

Produces a complete line of engineered wood products that have been certified by the Forest Stewardship Council.

Build with structural insulated panels (SIPs)

Green $pecs

Overall Rating

$ $ $

Difficulty

Green Benefits

ERIC SAYS The industry standard method of construction—wood framing—has been around for more than 175 years and has several steps. First, framing must be assembled with wood studs spaced every 16 (or 24) in. The studs have no rigidity, so large sheets of plywood sheathing are nailed on the outside of the wall. On the inside, large sheets of drywall are added to provide a smooth surface. Finally, a wall is filled with insulation for energy efficiency. All of these connections and steps are prone to leakage, weakness, and contractor error. It's a complicated process that could be made easier through technology.

That is the idea behind the structural insulated panel (SIP). SIPs are prefabricated building panels used for walls, floors, and roofs. The panels consist of a sandwich of rigid insulation between two structural layers of plywood. Think of them as framing, sheathing, drywall, and insulation all in one package. The panels are screwed together to create a strong and—most importantly—energy-efficient building.

Initially, SIP panels were made from virgin plywood skins over oil-based Styrofoam™ insulation. Today, greener options are available, including panels containing 100% recycled expanded polystyrene between certified and formaldehyde-free wood.

SIP panels are produced in a factory rather than on a construction site, which means high-quality construction. The pieces can be ordered precut, further speeding up the construction process. The panels arrive on the construction site in numbered order and are assembled like a giant puzzle. All the improvements decrease labor, reduce waste, increase strength, and improve the energy efficiency of a home.

What will this project do for your home?

SIP panels provide numerous advantages over standard wood framing:

- Strength: In SIP construction, every wall is a structural wall, making them more resistant to hurricanes and earthquakes. Homes built from SIPs were reported to have survived the 1993 earthquake in Kobe, Japan—and Hurricane Ike, which struck Texas in 2008—while neighboring homes were destroyed.
- Energy efficiency: The thick, insulated walls of a SIP home—combined with the tight fit of its components—create an incredibly energy-efficient interior. Studies show a SIP house can have twice the heating and cooling efficiency of a home made without SIPs and be 15 times less leaky and drafty than a stick-built home.
- Lower maintenance: Fewer air leaks in a SIP home mean it requires less maintenance and fewer contractor callbacks.
- Increased comfort: The higher insulation values and fewer drafts in SIP construction create a more comfortable home.
- Air quality: Generally, SIPs do not emit any VOCs or other harmful chemicals. They are low-emitting materials that create a home with cleaner indoor air quality.

What will this project do for the Earth?

According to the Energy Information Administration, houses are responsible for more than 20% of carbon emissions and other greenhouse gases in the U.S. By reducing the amount of energy consumed in a home, the threat of global warming is directly reduced. A SIP-built home can save 60% off its heating and cooling bills compared with a traditional stick-built home.

Producing SIP panels requires half of the energy and wood needed to construct a wood-framed home, which reduces the rate of deforestation and helps slow the destruction of global forests. SIP construction also greatly reduces construction waste. While building a typical home generates 8,000 lb. of debris, a SIP-built home will cut that by 60% and lower the amount of waste that ends up in crowded landfills.

Will you need a contractor to complete this project?

Building with structural insulated panels requires a bit of a learning curve. Although the techniques of

Structural insulated panels (SIPs) are used to build the exterior walls, floors, and roof of a home. The panels arrive precut and are highly energy efficient.

basic carpentry are used, the details of how the panels come together demand an understanding of the specifics. Search out a contractor who has experience using SIPs. For contractors unfamiliar with SIP construction, experienced subcontractors can be hired to complete that portion of the work. The Structural Insulated Panel Association (www. sips.org) has a list of certified installers on its website.

What are the best sources for materials?

A typical SIP panel is 4 ft. wide (the width of a typical sheet of plywood) and available in lengths of 8 ft. up to 24 ft. Construction goes much faster if you take advantage of a SIP's 4-ft. width, so push for the design of the building to be in increments of 4 ft.

With a solid core of insulation, SIPs provide superior energy efficiency. By comparison, if a typical wood wall offers an insulating value of R-11 to R-19, a SIP wall provides R-15 to R-45. The thicker the panels, the more insulation they provide. Choose the thickest panels you can afford and you'll achieve dramatic savings in your energy bills. With

A SIP is a sandwich of foam insulation between two rigid sheets of plywood. Hollow cores are often predrilled through the panels to allow space for plumbing and wiring.

standard thicknesses of 4 in., 6 in., 8 in., and 10 in., each step up translates into a nearly 20% drop in heating and cooling bills.

Although an 8-ft.-high SIP panel can be lifted by two people, anything longer will require the use of a crane. Panels weigh about 3 lb. per sq. ft. The panels are quickly tilted up into place and screwed together. A piece of solid wood is fitted between panels as a connector. The same panels used for the walls can also be used for the floor framing and roof.

The rectangular shape of SIPs creates boxy buildings without curves, but SIPs connect easily to other building materials. In fact, SIP walls are only used as exterior walls to support a building. Interior walls are nonstructural and can be constructed from any material you choose.

Once SIP walls are in place, they are treated like standard wood-framed walls and covered with siding. SIPs accept any siding material available, including stucco, shingles, and boards. Some SIPs are available with fiber cement siding already installed. While this speeds up construction, it also limits your options.

There are dozens of SIP manufacturers that offer a wide range of materials and options. It can be difficult to choose among SIPs, so look for ones with the following qualifications:

- Made with wood certified by the Forest Stewardship Council
- Use plywood with no added formaldehyde
- Made with zero-VOC adhesives
- Filled with recycled foam insulation
- Use encapsulated polystyrene (EPS) instead of polyurethane

If you still cannot decide on a manufacturer, select the one located as close to the project site as possible. You'll save on shipping costs and reduce the energy required to ship the panels.

Because SIPs are cut so precisely, the foundation must be accurately set. Any changes made to the foundation on site will create problems once the panels arrive. Changing panels on site is difficult and expensive. After the SIPs go in, set aside any cut scraps for possible reuse. The remaining pieces can be sent back to the manufacturer for recycling.

How much maintenance will be required after installation?

SIP panels do not shrink, warp, or twist like solid wood. The tight fit of the components creates a strong and well-built home. Maintenance of the exterior of a home will be much less than you'd find with a traditional wood-framed house.

SIP panels are also strong enough to allow for second-story additions, so building with SIPs offers the possibility of future renovations to a home without major structural changes.

How long will the project take to accomplish?

Planning the construction of a SIP home requires extra time early in the process. Finalizing the panel sizes, seeking out a trained contractor, and getting building permits approved adds a few weeks to the process. With an experienced contractor, this time can be easily made up during construction. A SIP-built home goes up quickly once the foundations are complete.

RESOURCES

Structural Insulated Panel Association
www.sips.org

KEVIN SAYS Structural insulated panels are one of several building alternatives to so-called stick building that seem to make sense. Why build things from scratch by hand when you can monitor construction and quality in a controlled environment? Sometimes we have a tendency to want things done the old-fashioned way when dealing with something as large and important as a home. But advancements in SIP technology have steadily increased its popularity.

What is the capital cost?

SIPs alone can be a little more expensive than the raw materials needed to construct a similar wood-frame home. Estimates run from $1 to $3 per sq. ft. in material costs. So a 2,000-sq.-ft. SIP home would require an additional $2,000 to $6,000 for materials. Most builders agree, however, that the material cost is compensated for by the reduced need for labor since SIPs are preassembled and install quickly. Most contractors surveyed agree that the net cost for SIP construction is equal to that of stick-built homes.

What financial resources are available?

There aren't any specific tax credits or rebates for the use of SIPs.

What is the monthly cost or savings? $

There have been documented savings on cooling and heating bills with SIPs over uninsulated construction, but similar savings can be achieved in a wood-frame home with efficient insulation (see p. 123).

What is the long-term home value?

I checked with several appraisers who told me there is insufficient data to support the idea that buyers will pay a premium for homes built with SIP panels.

THE BOTTOM LINE is...

SIP dealers like to tout energy savings as a primary motivator for their products, but it's really the reduction of waste that best helps the planet. It won't cost you any more than traditional wood framing, so if SIPs appeal to you, then have at it. ■

Build with insulating concrete forms (ICFs)

ERIC SAYS Although 90% of homes built worldwide use wood-frame construction, it's a system full of maintenance, durability, and strength issues. Wood framing is light in weight, but it is not very strong on its own. Heavy sheets of plywood and metal connectors must be added to make a wood house function in hurricane- and earthquake-prone areas of the country such as Florida and California.

The issues with wood don't end there. Extensive steps must be taken to protect wood buildings from fire. Hollow wood walls also have no insulation value, so insulation must be added to make a home energy efficient. A leaky roof can threaten the structural integrity of the walls, and if you're not careful, termites can infest the structure of the house.

Concrete construction offers an attractive alternative to wood. The unprecedented strength of concrete resists hurricanes, earthquakes, and fire. The material has the ability to store heat and cold, and this thermal mass keeps indoor temperatures more consistent. The heaviest rainstorm won't leak through concrete, and termites don't like the taste.

But pouring concrete is labor intensive, requiring complex formwork molds and skilled workers. Premade concrete blocks are available, but they are heavy and expensive to ship. Although concrete has a high thermal mass, it has a small insulation value and is unable to prevent the transfer of heat and cold.

In the 1960s, several clever engineers got together to combine the best features of both wood and concrete. They developed the insulating concrete form (ICF) to combine the strength of concrete with the lightweight energy efficiency of wood construction. An ICF consists of a hollow block made from lightweight foam insulating material that provides much-needed insulation for a wall. ICFs function as a mold (or formwork) for the concrete that remains in place to form the wall.

Green $pecs

Overall Rating

Difficulty

Green Benefits

Concrete is poured into the hollow center to create a solid wall. The system is quickly assembled, and the resulting wall is incredibly strong and energy efficient.

What will this project do for your home?

ICFs offer three distinct advantages over standard wood construction: energy savings, strength, and mass. The potential energy savings are impressive. Compared with a typical wood-framed home, a house built out of ICFs can be expected to cut 30% to 40% off heating and cooling needs.

The difference in strength is literally the difference between wood and concrete. ICF homes perform better in the high winds of tornadoes and hurricanes and during the rumblings of earthquakes. Plus, since the walls are solid concrete, there are fewer drafts, leaks, or cold spots— making a home more comfortable. Owners of ICF homes also notice how remarkably quiet their houses have become due to those thick, massive walls. And last but not least, the threats of mold, termites, and water damage are not a concern with concrete walls.

Concrete also has the ability to store heat and cold through thermal mass, which maintains a more consistent interior temperature and resists large temperature swings throughout the day. As sunlight strikes concrete during the day, it stores heat. When the temperature drops in the evenings, the walls release this stored heat into the home.

What will this project do for the Earth?

Heating and cooling consumes anywhere from 50% to 70% of the total energy used in a home. In the U.S., it accounts for 356 billion kWh hours of electricity used every year. If half of this energy could be saved by using energy-efficient ICFs, it would cut nearly 141 million tons of carbon dioxide from the atmosphere, which is the equivalent of removing the emissions of 23.4 million cars off the road, or the annual pollution from 30 coal-fired power plants.

Avoiding the use of wood in our homes would also save nearly 2 million acres of forest each year and would further slow the rate of global warming.

Will you need a contractor?

Building with ICFs is not quite as simple as stacking blocks, but it's close. Using ICFs is less laborious than building with standard concrete

ICF Wall Inner Web

An ICF is typically made from a shell of thick foam with plastic or metal webs inside. The forms are stacked to make exterior walls and can be cut with a simple handsaw during construction. Reinforcing steel bars (rebar) are set inside the hollow cavity, in the middle of the blocks, to add strength. Concrete is pumped into the walls in 3-ft.-high layers to keep the mix consistent. Concrete cures in seven days (reaching full strength after 28 days), and the ICF is left in place to act as insulation.

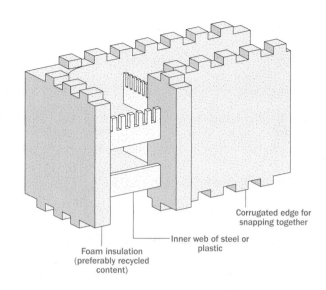

Corrugated edge for snapping together

Inner web of steel or plastic

Foam insulation (preferably recycled content)

blocks but requires knowing how to mix and pour concrete. Look for a contractor experienced in using ICFs.

Window and door openings also need to be planned carefully. Forgetting to install a window opening can be corrected later but requires expensive jackhammering. Concrete is not as forgiving as wood framing, so the method requires slightly more on-site preparation.

What are the best sources for materials?

Insulated concrete forms are widely available from dozens of manufacturers around the country. Their use is suggested for most parts of the country, and most local building codes permit them. While it may be inappropriate to build a concrete home in places with extreme cold (Alaska) or extreme heat (Las Vegas), the insulation of ICFs makes them suitable in nearly any part of the world. When choosing ICFs, select the thickest blocks you can afford. Thicker walls translate into greater energy efficiency and increased comfort.

Building with ICFs requires initial planning, but they can be used to create homes in any style—from traditional to modern. Building with ICFs is a lot like building out of Lego® blocks, as both quickly snap together to form the outline of a wall. Small ICF blocks can be set to create gentle curves or sharp angles, and their strength allows for large window openings and any type of roof.

ICF blocks can be made from a heavy mix of recycled materials and concrete. The edges of the blocks can be cut to create curves or rounded corners.

The foam walls of an ICF block are predominately available in two materials: expanded polystyrene (EPS) or extruded polystyrene (XPS). Both are similar to Styrofoam™. EPS and XPS are good insulators, but they are also oil-based plastic foams and not the greenest choice for a home. There are ICFs available made from recycled milk jugs using 100% recycled high-density polyethylene (HDPE), so be sure to look for companies using high recycled content for the foam and center web supports (see Resources on p. 256).

The construction of an ICF home begins like most homes, with a solid foundation. ICF blocks can be used over a traditional concrete foundation. But if you're planning on building an insulated basement, ICFs can also be used to construct the walls of the foundation itself. The recent surge in popularity of ICF blocks has also led to clever innovations like interlocking systems for the blocks. Jagged teeth are cut along the top and bottom edges of the blocks that allow them to snap together. Designed to eliminate the need for glue or ties, blocks with interlocking teeth speed up assembly and are highly recommended.

An ICF home is nearly airtight and will require additional ventilation. Since the finished home will be quite energy efficient, you will likely be able to use a smaller heating and air-conditioning unit. Discuss the appropriate size of units with your architect and engineer. They will be able to adjust their plans accordingly.

How much maintenance will be required after installation?

The maintenance concerns associated with a typical wood home disappear when using ICFs. Mold, mildew, wood rot, termites, carpenter ants, and rust are all eliminated. While you'll still have the same maintenance required for the stucco or siding, these are on the exterior surface and easily accessed.

Choosing ICFs

Search out ICF manufacturers whose products have the following features:

- Foam that contains zero VOCs or has zero off-gassing
- High recycled-content materials for the foam, plastic ties, and/or metal webs
- Blocks that do not contain any brominated flame retardants (natural borates for insect prevention are acceptable)
- Produced locally, within 500 miles of your home
- Accepts the exterior finish (stucco, siding, shingles) you plan to use

How long will the project take to accomplish?

The key to a successful ICF project is planning. Concrete is not as forgiving as wood, so the placement of ICF blocks needs to be verified carefully before pouring. Although there is an initial learning curve in using ICFs, the labor savings quickly appear. Walls that once took weeks to build can be completed in days.

In addition, insulated blocks allow concrete to be poured during more extreme weather. The ability to pour concrete during the coldest winter or hottest summer days reduces project delays and speeds construction. Concrete pouring often requires special inspections from the local building department, so adjust scheduling ahead of time.

KEVIN SAYS The arguments for ICF construction are mostly centered on its insulating qualities. But the process itself is not the epitome of planet-saving green. Even with fly ash added, Portland cement has environmental impact. And should a house someday be demolished, the polystyrene and cement pose a disposal problem. Many critics of ICF building feel the energy savings are not worth the environmental risks associated with the process.

ABOVE Large piping in a house must be installed before pouring concrete into ICFs. Smaller plumbing and electrical runs can be installed by cutting away the foam after the concrete has cured.

LEFT Once ICF blocks are stacked and the reinforcing steel has been installed, the concrete is poured, usually in 3-ft.-high intervals, which keeps the concrete from being too heavy at the bottom.

Concrete's Cost

Concrete is a durable and natural material, but its production takes a toll on the environment. The chief ingredient in concrete is Portland cement, responsible for some 110 million metric tons of global-warming emissions each year. However, it turns out that the waste soot from power plants, called fly ash, makes a comparable replacement for Portland cement. By substituting up to half of the Portland cement in concrete with fly ash, the environmental impact of the concrete inside ICF walls is drastically reduced. ICFs should really only be used if a high volume of fly ash concrete is also used in the construction. See p. 215 for more information on fly ash and concrete.

RESOURCES

Durisol®
www.durisolbuild.com

Makes ICFs from recycled wood chips mixed with concrete. Blocks are 12 in. high, not the standard 8 in., but install like traditional ICF blocks.

NUDURA®
www.nudura.com

Produces ICFs made from 100% recycled HDPE with webs formed from 100% recycled steel.

RASTRA
www.rastra.com

Made the first commercially available ICFs. RASTRA ICFs do not come in blocks but in long, thin panels up to 10 ft. The panels are available in 15-in. or 30-in. heights and can be installed horizontally or vertically. The narrow panels are a mix of recycled expanded polystyrene and cement, so they are heavier than traditional ICFs. The added cement allows exterior stucco to be applied directly onto the wall.

What is the capital cost? $$$$

Material costs for ICFs tend to run a little more than the material costs for stick-built homes— estimates run anywhere from 5% to 10% in additional materials and labor. For a conventionally built home costing $300,000, this could add an additional $15,000 to $30,000.

What financial resources are available?

We weren't able to dig up any specific tax credits or rebates for the use of ICFs.

What is the monthly cost or savings? $$

ICF homes definitely require less energy to heat and cool than conventional stick-built homes. However, most of this energy savings can be equaled in a wood-frame home with good insulation. One might still see as much as an additional 5% savings with ICFs depending upon the climate, which might account for a $15 to $25 savings on a typical monthly energy bill. In addition, there can be insurance benefits for the structural soundness of concrete construction in places like California and Louisiana, where natural disasters are the norm.

What is the long-term home value?

According to several appraisers interviewed from various parts of the country, there isn't enough data to support any long-term value impact associated with building with ICFs.

THE BOTTOM LINE is...

With the additional costs and environmental risks, ICFs are a less-attractive method for building than SIPs or even well-insulated stick-built construction. If you love the strength of concrete, you may consider ICFs. Otherwise, save your money and find other ways to help the planet. ▪

Install radiant heat

ERIC SAYS Every home has a heating system. In most new homes, heat is provided by a central system in which a large heater pumps hot air to each room through a network of ducts. Central heating systems tend to be noisy, carry dust and allergens through ducts and around the house, and can heat inconsistently, leaving one room stuffy and another room cool. Central systems are also not very energy efficient—they have to heat up all the air in a home even though only a few rooms are in use, and reheat air frequently to maintain a consistent temperature.

Older homes typically use wall radiators to stay warm. Rather than heating the air, wall heaters use gas or steam to release heat (through radiation). While efficient, they only really keep the immediate area warm. The heat is not distributed throughout the space.

An alternative heating system—radiant heating—warms a home via plastic tubes or electric wires installed beneath a floor. Heat rises, slowly and evenly heats the floor surface, and rises naturally to warm occupants—not all of the air in an entire room. The floor mass is used to maintain a consistent temperature, which can cut energy use by 20% to 40% over traditional heating systems. Individual rooms, or zones, can be heated independently, unlike with central systems. Dust and airborne pathogens are no longer a problem since heat isn't traveling through the air, and heat is much more evenly distributed than with wall radiator systems.

What will this project do for your home?

Radiant heat is considered by many architects and builders to be the most comfortable, healthiest, and energy-efficient heating method available. The virtually silent system does not spread germs and dust around the house and greatly improves indoor air quality. The effi-

Green $pecs

Overall Rating

Difficulty

Green Benefits

Comparing Radiant Heat with a Forced-Air System

With typical forced-air heating, hot spots are created at the heater and in high-ceiling areas. Radiant heating provides a more even temperature throughout the house, increasing comfort and reducing airborne-dust issues.

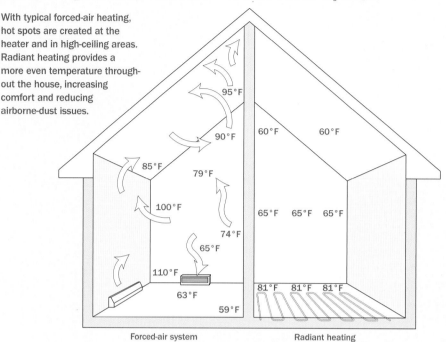

95°F
90°F
85°F
79°F
100°F
74°F
65°F
110°F
63°F
59°F

60°F   60°F
65°F 65°F 65°F
81°F 81°F 81°F

Forced-air system   Radiant heating

ciency of radiant heat can cut the heating bills of a house with a traditional forced-air system by up to half. In addition, because radiant heat slowly maintains a constant temperature, heat loss is cut by 25%. Radiant heat is also much safer than other heating systems, with greatly reduced risk of gas fumes and fire, and no dangerously hot radiators or dirt-collecting open vents.

What will this project do for the Earth?

About two-thirds of homes in the U.S. use oil, liquid petroleum, or natural gas for heating. These fossil fuels are in short supply and expected to run out by 2030, but our homes continue to consume approximately 22% of all natural gas used in the country.

One-third of U.S. homes use electricity to provide heating. Switching to radiant heating could save enough electricity to power 2.3 million homes a year and eliminate the emissions of more than 3.2 million passenger vehicles or almost four coal-fired power plants.

In total, heating our homes emits 150 million tons of carbon dioxide into the atmosphere each year, accelerating the risks of global climate change. By switching to energy-efficient radiant heating, emissions are cut and the effects of global warming are slowed.

Will you need a contractor?

There are two basic types of radiant heating: hydronic and electric. Hydronic systems require a complex installation of tubes, boilers, and plumbing connections. Hydronic systems can be installed into existing homes only if replacing existing floors or if there is an accessible crawlspace below the floor. They are impractical to install over existing floors. The added weight of a hydronic system must also be factored in—you must confirm that an existing floor can support it. Use only an experienced, professional installer for hydronic installations.

An electric radiant heating system consists of a thin wire mesh installed below finished flooring. Its tiny size makes it ideal for use during remodels or in tight spaces, under tile, carpet, or even wood flooring. Although it uses low-voltage wiring, electric radiant heating still requires an electric connection. Turn off all power before starting any work. Consult a professional electrician if you feel in over your head.

What are the best sources for materials?

Electric systems are not as energy efficient as hydronic systems, but they are ideal for bathrooms, where heat needs to be on only while the room is in use; beneath tile floors, where electric mesh fits easily below the tiles; in solar-powered homes because electric radiant systems can be operated on "free" solar energy; and in existing homes where floors cannot fit or support hydronic tubes. Unless there are solar panels on your roof, the cost of operating electric radiant throughout your home might not make much sense.

Electric mesh must be handled carefully during installation to avoid cutting the electric mats and breaking the circuit. Test the system throughout installation to ensure it is working. Electric radiant heating can be purchased from home improvement stores or local tile or carpet showrooms.

The majority of radiant heating systems are hydronic systems, where hot water (hydro) is circulated through a network of cross-linked polyethylene tubes embedded below the surface of a floor. The temperature of the water in the tube only needs to be heated between 90°F and 120°F to function.

Hydronic systems are ideal for new homes, where floors can be designed to accommodate the weight and thickness of the tubes (usually about $1/2$ in. thick); in homes made with concrete slabs, as

Radiant heating can be installed over wooden floor structures using special underlayment made from plywood sheets with precut grooves specifically designed for the tubing.

the concrete mass retains heat well; in homes with a solar thermal system, because a hydronic radiant system would be "free" to operate; and in existing homes where there is a basement or crawlspace access and tubes can be stapled to the underside of floor joists.

Hydronic radiant heating is also commonly installed into a concrete slab. Tubes are placed onto the floor before the concrete slab is poured. Once poured, the concrete completely covers the tubes. Concrete is the ideal material for radiant heat, as the mass of the slab stores heat and requires a minimal amount of heat to maintain a constant temperature. Even after a radiant system is switched off, the concrete mass continues to radiate stored heat into the home.

For homes with wood-framed floors and crawlspaces, radiant tubes can be installed into the floor using radiant panels that are put in place of a plywood subfloor and covered with a thin metal plate that helps radiate heat upward into a room. The panels speed up installation and improve the performance of a radiant system by 10% to 20%.

Although hydronic radiant heat can be used with nearly any type of flooring, certain materials work better than others. Heavy-mass materials, such as concrete, stone, tile, and brick, are ideal for radiant heat, as they conduct and store heat well. Carpeting can be used, although it acts as a layer of insulation between the tubes and your feet. Wood flooring can be used as long as you allow the floor to adjust to the humidity of the house for a couple of weeks before installation.

The water in a radiant system is heated in a large hot water tank called a boiler, which is often placed alongside a home's standard domestic hot water heater. Special all-in-one, dual-purpose boilers are available that heat water for household use and a radiant system in the same tank, but on-demand (or tankless) water heaters (see p. 141) cannot be used with radiant heating systems.

Radiant heating works in a different way than forced-air systems. Radiant heat slowly warms the mass of an entire building to keep you warm. It takes several hours for a building to heat up and cool down. In places with unpredictable climates that can be cold on Monday but warm on Wednesday, radiant heat might not be suitable. But it is ideal for climates with long periods of cold winter. Hydronic and electric systems can also be used together to warm a home.

How much maintenance will be required after installation?

Hydronic radiant systems require a simple annual check of the water-pressure gauge. A drop in water pressure is an instant indicator of leaky tubing. Electric radiant systems have no moving parts and require virtually no maintenance. Both systems have a life expectancy of 30 to 40 years.

How long will the project take to accomplish?

Electric radiant mesh installs seamlessly during a normal flooring installation, adding only a couple of hours per room to an overall project. The wiring for the control switch should be put in place before walls have been completed. The wall switches can be added to an existing room with a little more effort.

Hydronic installation time will vary depending on the type of flooring, the number of zones, and the size of the house. The installation schedule will have to be coordinated with other trades, such as the flooring installers. In general, adding a couple of weeks to the schedule is typically enough to allow for the installation, but confirm timing with your builder.

KEVIN SAYS Imagine stepping onto a bathroom floor on a cold winter morning and amazingly your feet don't feel like blocks of ice. This is the major comfort attraction for radiant heating systems, along with the lack of noise and (questionable) efficiency. Despite its cachet, there is debate over the savings-versus-cost aspect. Many in the green community point to super-efficient homes that don't require much heating, significantly reducing the savings benefits of radiant heat.

What is the capital cost? $$$$$

There are a few factors driving the additional cost of radiant heat versus forced-air or baseboard heating. First, the labor for installation is much greater with radiant heat.

To install radiant heating in a home where existing floors cannot be disturbed, tubes can be stapled to the underside of flooring joists, which requires basement or crawlspace access to reach the floor above. This method is not as efficient as a standard radiant system and often requires hotter water temperatures to radiate to the floor above. Aluminum transfer plates can be added to increase efficiency, but they make installation slightly more involved.

Installation can run from $6 to $12 per ft., totaling $12,000 to $24,000 for a 2,000-sq.-ft. home, compared with $2,000 to $5,000 for ducting on a forced-air system or $1,500 for floorboard heaters. The second cost factor is the need for additional ducting for air-conditioning. Most forced-air systems today use the same ducting for both, which makes radiant heating a separate expense on its own. Lastly, radiant boiler systems cost $1,000 to $2,000 more than conventional forced-air units.

What financial resources are available?

There aren't any specific tax credits or rebates for the use of radiant heat that we could find.

What is the monthly cost or savings? $$$

Radiant floor systems are about 10% to 25% more energy efficient than conventional systems, but climate and personal habits impact overall energy usage. In a 2001 Canada Mortgage and Housing Corporation study, thermostat settings in radiant-heated homes averaged more than a degree higher in usage than homes with conventional heating systems, wiping away much of the savings to be had. Figuring $15 to $40 a month in savings on a like-for-like basis requires 20+ years to make up the initial cost. Additional savings come from the durability and low-maintenance aspects of the system since radiant heating can last 10 to 20 years longer than conventional heating systems.

What is the long-term home value? $

Many buyers will be lured by the convenience and luxury of radiant heat, but don't expect to make your money back. Several appraisers confirmed that there's no evidence to support the theory that people will add the cost of radiant heat to the resale price of a home.

THE BOTTOM LINE is...

Radiant heat certainly has plenty going for it in comfort and high-end reputation, but economics are not the major selling point for the system. There is enough energy efficiency data to support making this part of a green home approach, but don't look to recoup dollars spent. ■

Integrate your energy systems

ERIC SAYS Powering lights, watching television, heating water for the shower—all of these functions require our homes to use an incredible amount of energy. Buildings are the biggest global source of energy consumption, surpassing the usage of automobiles by almost twice as much.

Using a method called "building integrated solar," electricity, heat, and hot water can all be generated from the sunlight hitting a home. Instead of using three separate systems, building integrated solar often involves the careful combining of solar panels, solar thermal, and radiant heating into one shared system.

Building integration takes advantage of the overlapping functions of each separate system to make a single, more energy-efficient one. The integration of energy systems enables them to work together for added efficiencies and energy savings.

What will this project do for your home?

Almost 20% of energy used at home goes toward water heating, another 40% goes toward electrical needs such as lighting, appliances, and air-conditioning, and the remaining 40% is used for heating. All three areas are affected by using building integrated solar and can reduce overall energy use by more than half.

Building integrated solar also means savings from purchasing less equipment and sharing equipment and resources. And—as a by-product—the system produces free hot water.

What will this project do for the Earth?

The use of solar panels and solar hot water heaters requires no energy to operate, nor does it produce emissions. Residential energy use can

be slashed in half, saving millions of pounds of carbon emissions and avoiding the use of coal, natural gas, and other fossil fuels.

By producing power and heat with building integrated solar, our dependency on fossil fuels is lowered and the risks of global warming are greatly reduced.

Will you need a contractor?

If building a new home, select an integrated system during the design process so that the home is able to accommodate the weight and size of the necessary equipment. Solar panels and heaters are best installed by professionals.

Not every solar power provider deals in integrated solar products. Explain that you are looking to use a combined system for producing both heat and electricity. Check references and only use a company that is licensed and bonded to do the work. Any good solar installer will need either a copy of the floor plans or a site visit to determine the size and cost of the system. The orientation and amount of sunlight will determine the size and location of the equipment. You'll need a sunny area of approximately 200 sq. ft. on your roof.

What are the best sources for materials?

The technologies of integrated energy systems fall into three main categories:

Building integrated photovoltaics (BIPV) As the name implies, BIPV conceals solar panels into the overall design of a building. The most common BIPV product is solar panels that double as roofing shingles. You can mix and match solar shingles with standard ones. Solar shingles (sometimes called "solar slates") do not produce as much energy as standard solar panels of the same size but are ideal in places where the panels need to be concealed or camouflaged. What is lost in efficiency is gained by having one product serve as both the roof and solar panels. In addition, the enhanced appearance of solar shingles should meet any requirements from a homeowners association or architectural review board. Other BIPVs include overhangs or skylights made from solar panels.

Solar shingles replace real shingles and blend seamlessly into the roof. Although less efficient than traditional solar panels, they are great in places where solar panels need to be discreet or hidden.

Building integrated solar thermal (BIST) BIST consists of plastic tubing installed on the outside of a house to generate hot water. Most BIST installations are custom assembled from standard radiant-heating components. The hot water can be used for showers or as a heating source. The tubes are placed in rows on a sunny area of a wall or roof and covered with siding or roofing. Dark-colored metal panels or shingles are typically used, as metal conducts heat efficiently. Water is pumped through the tubes to heat the water, where it is stored in a standard water heater tank.

Combined solar PV and thermal panels Typical (electricity-producing) solar panels get very hot sitting on a roof. In fact, 60% of the sun's energy hitting the roof is lost as heat. A combined solar PV and thermal system collects that leftover heat to produce hot water or hot air. The panels are identical in size and appearance to typical solar panels but don't require two separate products to produce electricity and heat. The systems can also be used over any type of roofing material.

Integrated solar systems can be used throughout North America, and since they produce heat, they are actually well suited for colder climates. Sunny and cold areas work just as well as sunny and hot places. But the systems won't be as effective in cloudy regions.

How much maintenance will be required after installation?

Building integrated solar requires minimal maintenance. The panels should be hosed off occasionally to work efficiently. Use a power washer to knock the grime off the panels once a month.

BIST systems must be checked monthly for leaks. A simple pressure gauge is all you need—a drop in pressure indicates leakage.

Building Integrated Solar Thermal

Building integrated solar thermal (BIST) usually involves installing a large panel on a south-facing wall or roof that pulls air in, heats it using sunlight, and recirculates the heated air back into an interior room. A solar panel provides all of the electricity required for operation.

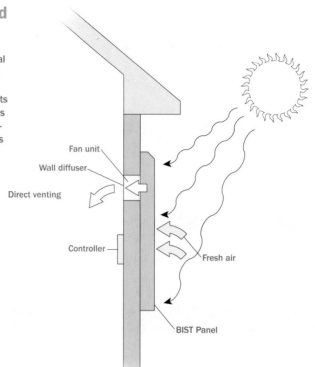

Fan unit

Wall diffuser

Direct venting

Controller

Fresh air

BIST Panel

How long will the project take to accomplish?

Once the products and installer are chosen, installation can often be completed in a couple of weeks. Order materials ahead of time to prevent construction delays. Unlike standard solar panels, most BIPV and BIST systems cannot be installed once the roof and walls are completed.

KEVIN SAYS Integrating systems makes great economic sense from an installation and financial point of view. The benefits of solar are increasing, and taking a one-step approach offers a chance to consolidate space and preserve the aesthetics of a home from an unsightly mishmash of pipes, tubes, and panels. Economies of scale can be had as well, not to mention the convenience of only having to put up with noisy roof construction once.

What is the capital cost? $$$$$

Integrated solar systems aren't cheap and can run anywhere from $15,000 to $75,000, depending on the size and needs of a home. However, there are savings of 20% to 40% over purchasing and installing individual solar systems for water, heat, and energy.

What financial resources are available? $$$$$

There is plenty of government money through states and municipalities to alleviate upwards of 50% of solar cost through grants, low-interest loans, and tax credits. For more information, visit the Database of State Incentives for Renewable Energy (DSIRE) website at www.dsireusa.org.

What is the monthly cost or savings? $$$$

The combined energy savings of an integrated system may be as high as 50% of energy bills, depending on usage and climate, which could amount to $1,800 annually on a $300 monthly energy bill in a typical home. That's likely to be enough to pay for the combined system in as little as 8 to 10 years after rebates and tax breaks.

What is the long-term home value? $$$

The combined energy-efficiency aspects of an integrated system will appeal to green-conscious buyers. As with independent solar units, it's unlikely you'll get the full cost in a resale, but as long as the system is not unsightly, you may recoup as much as 20% to 30% of your investment.

THE BOTTOM LINE is...

If you are spending the money to go solar, then you might as well get as much bang for your buck as possible. Combining systems gives you aggregated savings in installation and overall savings on energy bills without extra punishment for aesthetic issues. ■

RESOURCES

Database of State Incentives for Renewable Energy
www.dsireusa.org

Solar shingles:

Atlantis Energy
www.atlantisenergy.com

Schott Solar
www.us.schott.com

Produces see-through solar panels that are used in place of skylights. Once installed, they resemble tinted glass.

SunPower® Corp.
www.sunpowercorp.com

Uni-Solar®
www.uni-solar.com

BIST systems:

Dawn Solar
www.dawnsolar.com

Combination units:

Bright Phase Energy
www.brightphaseenergy.com

PVT Solar
www.pvtsolar.com

Install a green roof

Green $pecs

Overall Rating

Difficulty

Green Benefits

ERIC SAYS In many homes, roofs are a bit of an afterthought. Their design is often the result of what is going on inside the house, rather than shaped by the direction of the sun or a desire to stay cool. But the design of a roof can play an integral part in the overall energy efficiency and sustainability of a home.

Most roofing consists of unsightly asphalt shingles or a flat membrane in a limited number of colors. But imagine a green roof covered with living plants that protects the roof membrane and adds insulation and beauty to the top of a house.

A green roof system does not replace the roof but is installed over a real roof. It's not simply setting potted plants up there, either, but includes soil, drainage, mulch, plants, and (sometimes) irrigation. A quarter of the heating and cooling in a home escapes through the roof. While adding attic or roof insulation can help (see p. 123), a green roof provides up to triple the amount of insulation of a typical roof.

What will this project do for your home?

Installing a green roof on a home improves indoor comfort, reduces heat gains in summer by up to 95%, and cuts heat losses in winter by more than 25%. A green roof reduces the need to run heating and air-conditioning units, resulting in 25% to 40% energy savings. Because less energy is needed, you could also install a smaller heating and air-conditioning system than would otherwise be necessary.

Since a green roof sits over a waterproof roof layer, it can double the life expectancy of a roof and help prevent leaks. The mass of the soil overhead allows a house to maintain a more consistent temperature and helps muffle outside noise.

What will this project do for the Earth?

Installing plants on a roof has an immediate and direct effect on global warming. The plants absorb carbon dioxide (a greenhouse gas) and produce oxygen. They naturally filter air and rainwater and provide a living habitat for birds and other wildlife.

More than 258 billion kWh of electricity are used every year just to air-condition our homes. A green roof is considered a cool roof (see p. 153), lowering the temperature on a roof by up to 70°F in summer. The temperature around the building drops as well, reducing the need for air-conditioning.

Will you need a contractor?

Although green roofs are relatively easy to install, there are several factors to consider before starting. A landscape architect or contractor can help design and install a green roof system, selecting the appropriate plants and specifying the correct roofing and drainage. While you can install the roof yourself, you may want to leave it to the professionals. Working on a high roof is hot, dangerous, and runs the risk of damaging the waterproof roof layer during installation. Choose a landscape architect or contractor who has experience with green roofs.

What are the best sources for materials?

Green roofs are available in two main types: extensive and tray. Extensive green roofs are a series of built-up layers that create a fully planted roof. A waterproof membrane is installed as the actual roofing layer. Over that, a protective layer of insulation is added. Then a layer of drainage material is put in place to allow water to pass to the roof drains. A moisture mat helps slow the water's passage and creates a layer for the plant roots to grow. Finally, soil and plants are brought in.

A green roof tray system consists of plastic, metal, or fiber pans filled with soil and plants. The tray systems allow for the installation of more mature plants that have had time to grow on the ground. They install quickly and allow for easy access to any portion of the roof should a problem arise in the future. The trays are typically lighter and thinner than an extensive roof, so they do not provide the same amount of benefit of insulation or flexibility as an extensive green roof. Your local landscape architect can help source more natural materials, such as hemp and burlap, which biodegrade into the soil over time.

Regardless of the type of system you choose, there are several factors of green roofing to consider.

Maintaining a green roof involves planting, weeding, and fertilizing, which means homeowners must invest time to do it themselves or money to find a gardener willing to work on a roof.

- **Weight:** The weight of green roofs is calculated at the same weight as water, about 62 lb. per cu. ft. Tray systems weigh from 12 lb. to 40 lb. per sq. ft., depending on the thickness of the tray. Extensive roofs are often thicker and weigh more. A roof must be designed to support the extra weight. If you live in an area with heavy snowstorms, your roof is probably already designed to support such loads.
- **Plants:** Green roofs can be installed anywhere in the U.S., as long as appropriate plants are selected. Choose drought-tolerant or native-species plants. Not every green roof needs to be irrigated. If you choose the appropriate plants, they can survive with rainfall after taking root. Fire is a serious concern with any building, but especially with green roofs. Select plants that can withstand the hot conditions on a roof and are appropriate for your climate. An irrigation system helps alleviate the threat of fire, but regular trimming of overgrowth is the best way to reduce fire risk.
- **Drainage:** Rainwater must go somewhere, so the roof must be designed to drain properly. Consult with a landscape architect for the proper methods of draining a green roof.
- **Slope:** Green roofs are typically installed on low-sloping roofs, but they do not need to be flat. Steeply sloped roofs are often impractical for planting.

- **Solar panels:** Green roofs and solar panels can't occupy the same spot on a roof. But the two systems can coexist by placing them side by side. Be careful the plants don't shade the panels.
- **Location:** Ideally, you'd want to install a green roof in a spot that is visible, such as from an upper floor or high vantage point. If you can't see the plants, you're better off adding more insulation to the roof and installing solar panels.

Existing buildings can enjoy the benefits of a green roof, but weight and drainage are often the deciding factors. Confirm with an architect or structural engineer that your existing roof structure can support the extra weight. If not, you may have to beef up the support before installing the plants. If an extensive green roof is too heavy, explore lighter-weight tray systems instead.

How much maintenance will be required after installation?

Like a new garden, your green roof will require additional watering and mulching until the plants take root. After that, the maintenance is the same as for any garden. Weeding, fertilizing, and replacing any eroded soil will be necessary. Trays require less maintenance since they can be removed to access the roof and can be replaced one at a time. Maintaining a green roof means you'll have less maintenance on the roof itself.

How long will the project take to accomplish?

Once designed, a green roof can be installed as soon as the roof sheathing is complete. Installation time varies based on the size of the roof, but a typical home can be completed in a few weeks. The plants will take root over the next year.

KEVIN SAYS The idea of a roof garden is nothing new in urban dwelling environments like New York City, but making the entire roof itself a garden is a bit more extreme. The benefits are obvious from an insulation standpoint, but there are significant cost and potential upkeep issues to consider. Not all roof designs are func-

RESOURCES

American Hydrotech, Inc.
www.hydrotechusa.com

**American Wick Drain
Corporation**
www.americanwick.com

Barrett Company
www.barrettroofs.com

Building Logics, Inc.
www.buildinglogics.com

**The Garland
Company, Inc.**
www.garlandco.com

**Green Roof Resource
Portal**
www.greenroofs.com

GreenTech, Inc.
www.greentechitm.com

Roofscapes, Inc.
www.roofmeadow.com

Soprema USA, Inc.
www.soprema.us

Green tray roofing:

GreenGrid®
www.greengridroofs.com

GreenTech®
www.greentechitm.com

tional for green roofs. Unless your home has a flat or low-sloping roof, then a green roof is out of the question.

What is the capital cost? $$$$

The cost of a green roof varies based on who installs it and the complexity of the system. Prices, including installation, range from $8 to $25 per sq. ft. If you were to select a 20-ft. × 40-ft. section of a roof, you would look at a cost of $6,400 to $20,000, fully installed. This represents a 40% to 80% premium over conventional roofing.

What financial resources are available?

There do not appear to be any tax credits or rebates for residential green roofs, although commercial breaks exist in New York City and attempts at legislation have been made that may pass in the future.

What is the monthly cost or savings? $

Some insulating value comes from a green roof, but compared with a reflective, well-insulated roof, the savings are minimal. The durability of green roofs is documented to have nearly twice the life span of a standard shake or composite roof, which saves on repair and replacement costs down the line. Plant maintenance will require time on the part of the homeowner or a gardener comfortable on the roof.

What is the long-term home value?

There's not enough data to support the idea that people will pay more for a green roof. Appraisers expressed concern that if the roof was not aesthetically pleasing, it would devalue the property. Buyers looking specifically for a green roof may be willing to pay a premium, but there is no telling if a number of such buyers exist at this time.

THE BOTTOM LINE IS...

A green roof makes for a fascinating conversation piece and, for an isolated custom home, may prove to demonstrate a commitment to green. Most of the benefit is to the atmosphere rather than to the specific home, and will come at a financial premium that should be taken into consideration. ■

Utilize passive solar design

ERIC SAYS Before the advent of heating and air-conditioning, buildings were designed to take into account the location of the sun and the wind. Architects and builders understood that certain materials held in heat and cold and could be used to maintain a consistent temperature in a structure. This approach is called passive solar design.

Passive solar design is powered by a clean, abundant, and nonpolluting energy source: the sun. It doesn't require any special materials or equipment, just a little extra care in how a house is designed. It is not a specific product to buy or set of procedures to follow. It is a strategy to alter the overall design of a home so as not to require a heater or air-conditioner to run (or at least not to run as much). Passive solar design can keep your home within 5°F to 10°F of your normal comfort zone and can cut heating and cooling bills by 50% to 90%.

Passive solar techniques are easy to apply if designing a new home, as you can control the site of the house and location of the windows. However, existing houses can be adapted to passively collect or control sunlight. Passive solar designed homes also go hand in hand with many energy-efficient features, including solar panels (see p. 90), solar water heating (see p. 147), energy-efficient windows (see p. 111), radiant heating (see p. 257), and rainwater catchment (see p. 196).

What will this project do for your home?

A well-designed passive solar home has drastically reduced heating and cooling bills. If the techniques are carried out correctly, overall energy needs will drop and a thermostat may only need to be used during the hottest or coldest periods of the year, saving on energy bills and reducing wear and tear on heating and cooling equipment.

Green $pecs

Overall Rating

Difficulty

Green Benefits

What will this project do for the Earth?

Heating and cooling our homes produces the equivalent carbon emissions of 25 million passenger cars, or almost 30 coal-burning power plants. Every home could utilize the principles of passive solar design and easily cut energy use by at least one-third. The hundreds of thousands new homes built in the U.S. each year could be designed to cut energy use by even more, simply by using passive solar strategies, which could drop global warming back to 1999 levels.

Will you need a contractor?

The basic concepts of passive solar design have been around since mankind started construction, but they take time to master. An architect, designer, and engineer can provide experienced, expert advice on how to best utilize passive solar techniques.

What are the best sources for materials?

Passive solar design is based on five basic physics principles:

- **Hot air rises:** Air warmed by the sun rises, and cooler air rushes in to take its place. Placing vents or windows in high and low locations helps create cooling breezes in the home.

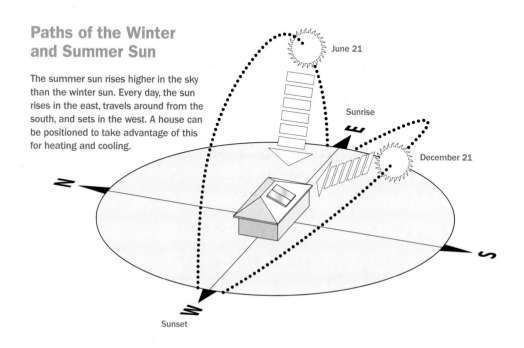

Paths of the Winter and Summer Sun

The summer sun rises higher in the sky than the winter sun. Every day, the sun rises in the east, travels around from the south, and sets in the west. A house can be positioned to take advantage of this for heating and cooling.

June 21

Sunrise

December 21

Sunset

- **High summer sun:** The sun is higher in the sky during the summertime. Well-positioned overhangs can shade the home and keep it cool during these times.
- **Low winter sun:** The sun is lower in the sky during the wintertime. Sunlight passes under overhangs to reach into the home and keep it warm. Don't make the overhangs too deep, or you risk blocking this warming winter sunlight.
- **Thermal mass:** Certain materials are able to hold in temperature better than others. Heavy-mass materials (such as concrete, stone, brick, or even water) store heat and cold well. When the temperature outside changes, these materials release stored warmth or cold to help regulate the heating or cooling in a house.
- **Insulation:** A well-insulated building maintains an interior temperature, despite changes to the outside temperature, better than a poorly insulated building.

Orientation to the sun If a home is designed to be only one room wide, it allows sunlight to penetrate the width of an entire house. The long side of the house should be oriented to run east to west to face south. This brings in abundant natural light but also allows you to control the amount of sunlight that comes into the home with overhangs and strategic window locations. If a house cannot be positioned completely facing south, anywhere within 20 degrees in either direction will still work.

Proper window placement In passive solar design, windows act as solar collectors—bringing in light and heat and allowing for ventilation. In hot climates, too many south-facing windows make a home too hot, while north-facing windows provide reflected light but no heat. In cold climates, you need south-facing windows to allow sunlight to warm things up during the day. To help hold heat in, thick curtains can be drawn on winter nights.

Using thermal mass Heavy or massive materials have a high thermal mass—the ability of a material to store temperature. For example, if you leave a brick in the hot sun all day, it will release the stored heat all night after the temperature drops. To apply this to passive solar design, a thick wall (at least 12 in. of stone, concrete, brick,

Strategies to Regulate Temperature

Keeping cool on a summer day. Large overhangs shade a house from the southern and western sun for most of the day. Inside, a thick concrete wall can absorb heat from the house and keep things cool.

Cooling off on a summer night. Open both high and low windows to get evening breezes flowing. This flushes a house with cool air, which is absorbed by a concrete wall all night and released to cool the house during the next day.

Warming up on a winter day. The low winter sun warms the south side of a house all day. A stone or concrete floor stores up the heat. Hot air is allowed to build up by the ceiling.

Keeping warm on a winter night. Stored heat in the floor and walls made from thermal-mass materials is released during the night to keep the house comfortable as the temperature outside drops.

etc.) needs to be positioned in front of a large window. The sun comes through the window and heats up the wall all day. Once the sun sets, the stored heat is released and warms the house.

The floor can also be used to store thermal mass. An exposed concrete slab at least 8 in. thick by a south-facing window will store heat all day. Leave about 10 ft. of exposed floor—not covered by carpet or wood flooring—around windows to take advantage of this.

Thermal mass is appropriate if you live in a climate that is warm by day and cool at night. Thermal mass requires this shift in temperature to function; it will not work in places that are consistently hot or cold.

Maintaining sun control Any windows located on the south side of a house should have an overhang 2 ft. to 3 ft. deep. Any deeper, and the low winter sun will not be able come through the window to keep the interior of the home warm. Any narrower and the unwanted, hot summer sun will come in relentlessly, baking the inside of a home. Windows along the west and east sides of the house should have deeper overhangs to shade the glare of morning and evening sunlight.

Overhangs can be made from awnings, trellises, or anything that shades a window. Deciduous trees can also be planted to provide shading. In the winter, these trees lose their leaves and the sunlight passes through the bare branches to warm the house.

House Orientation

Ideally, the length of the house is oriented to run along the east/west axis. This orientation, allows for the greatest control of the sun, heat, and light in summer and winter.

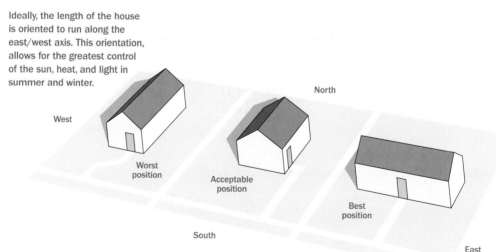

Positioning an Overhang

Overhangs can shade south-facing windows and control summer heat but allow winter sun to enter and warm the home.

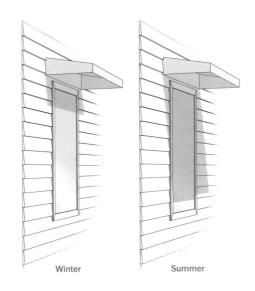

Winter   Summer

Creating ventilation Warm air is lighter than cold air, so it rises naturally. When hot air rises, cooler air rushes in to take its place. In the summer, hot air can be expelled from a home by installing high vents, clerestories, or operable windows. On the opposite side of the room, install low windows to allow cool air to enter the house. Open both the high and low windows and air will flow naturally through the house and keep it cool. In the winter, close all windows and trap the hot air inside to keep things comfortable.

There are a number of wonderful books and videos that get into the minutiae of passive solar design, but I prefer the free resources online such as the Sustainable Buildings Industry Council (www.sbi council.org), Build It Solar (www.builditsolar.com), and the Energy and Environmental Building Association (www.eeba.org).

How much maintenance will be required after installation?

Passive solar design principles are used to design and build a home. Beyond that, little specialized maintenance is involved with such a home—just normal home upkeep. The daily operation of a passive solar home means opening windows around the house on summer evenings and closing them the next morning. Although not difficult, it is a necessary part of making a passive solar home operate more efficiently.

Passive Solar According to Place

Hot summers, warm winters (Phoenix, AZ; Miami, FL): Avoid installing south-facing windows or skylights. If you include southerly windows, make them short with overhangs. Avoid using thermal mass on any exterior walls.

Hot summers, cold winters (Philadelphia, PA; Chicago, IL): Install large south-facing windows with standard overhangs that control summer heat but allow winter sun to enter. Avoid large windows on the north side. Place a thick thermal-mass wall close to a south-facing window to provide heat in the winter. Shelter the house by planting evergreen trees to the north.

Mild summers, mild winters (Portland, OR; San Francisco, CA): Thermal-mass walls can be installed on exterior walls to maintain a regulated interior temperature. Make most windows operable to allow for proper ventilation.

In humid climates, raise the house to get breezes moving through the crawlspace. Avoid planting shade trees nearby, as they stifle cooling breezes.

In dry climates, sink the house into the earth to increase cooling. Mounds of earth (called berms) can be built up against the house to further insulate.

How long will the project take to accomplish?

Incorporating passive solar design concepts into a home is simply a part of the design process. It should not add time to the project. The time needed to retrofit an existing home (by adding overhangs, windows, or thermal mass) varies based on the extent of the renovations.

KEVIN SAYS

When building a new home, it makes sense to create an environment that is as efficient and state-of-the-art as possible. Allowing the sun to do the maximum heating and lighting work in a home makes great energy sense and lends itself to creating a more comfortable living space.

What is the capital cost?

There is no inherent cost to taking a passive solar approach in a home design. Green architects can be found with affordable rates and the cost of labor and materials should be about the same.

What financial resources are available?

We couldn't find tax credits or rebates for houses made with passive solar design.

What is the monthly cost or savings? $$$$$

A home designed with a passive solar approach could cut 60% to 80% off heating and cooling bills versus a home designed without concern for the position of the sun. This could result in as much as a $120 to $240 savings relative to a $300 monthly bill. Additionally, less use of a heating and air-conditioning system means lower maintenance costs.

Passive Cooling with High and Low Windows

As hot air rises, it can be released through high windows to keep a building cool. Low windows provide a replacement of fresh, cool air.

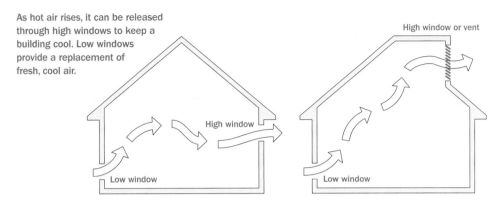

High window or vent

High window

Low window

Low window

What is the long-term home value? $$$

While there is no specific recorded data supporting a higher price paid for a passive solar home versus a comparable custom home, people will spend money for good design. Those attracted to an ecologically sensitive environment that benefits from the surroundings will likely be interested and may even pay a premium for an aesthetically pleasing, passive solar home when shown documented energy savings.

THE BOTTOM LINE is...

One of the great joys in building a custom home is to create a livable environment that is satisfying and efficient. The passive solar approach ensures that you get the most from nature as well as your dollar. It is an approach that will never lose its effectiveness and is, like a lot of other green principles, here to stay. So to maximize the home's green potential—you can start reaping the benefits now and be in a good position if and when you decide to sell down the line. ■

RESOURCES

Build It Solar
www.builditsolar.com

Energy and Environmental Building Association (EEBA)
www.eeba.org

Passive Solar Industry Council
www.psic.org

Sustainable Buildings Industry Council
www.sbicouncil.org.com

credits

index